Anarchy, Geography, Modernity: Selected Writings of Elisée Reclus

Edited and Translated by John Clark and Camille Martin

With an Introductory Essay by John Clark

Anarchy, Geography, Modernity: Selected Writings of Elisée Reclus Edited and translated by
John Clark and Camille Martin, with an introductory essay by John Clark

ISBN: 978–1–60486–429–8
Library of Congress Control Number: 2013911520

Cover: John Yates / www.stealworks.com
Interior design by briandesign

10 9 8 7 6 5 4 3 2 1

PM Press
PO Box 23912
Oakland, CA 94623
www.pmpress.org

Printed in the USA by the Employee Owners of Thomson-Shore in Dexter, Michigan.
www.thomsonshore.com

For all who are able to envision
a free, just, and compassionate world,
and who, like Reclus,
dedicate their lives
to creating such a world

Contents

L'HOMME EST LA NATURE
PRENANT CONSCIENCE D'ELLE-MÊME

Preface to the PM Press Edition

One of the best-known images from Reclus' works originally appeared above the preface of his magnum opus, *L'Homme et la Terre*, and is reproduced here. It depicts two hands holding the earth, coupled with the statement in French that "Humanity is nature becoming self-conscious." It is clear that the image indicates not only that the fate of the earth is now in the hands of humanity, but also that humanity can only fulfill its weighty responsibility by acting with an awareness that we are an integral part of nature, rather than continuing under the illusion that we are a power over and above the natural world. Reclus' message is that the "hands" in the image are those of nature acting through humanity, though it is up to the viewer whether to read the image with more emphasis on humanity or more on nature.

Another famous Reclusian image, the one reproduced on the cover of this book, contains no such ambiguity. In this image, we see Nature herself contemplating or watching over the earth, which this time is clearly held in her hands.[1] The contemplating and holding seem to be inseparable parts of one process. The image evokes aspects of the contemporary ethics of care, an important dimension of ecofeminism, in which "holding" is a key concept. Feminist philosopher Sara Ruddick introduced this idea to describe the maternal attitude of preserving, conserving, and maintaining what is needed in a child's life. Quoting Adrienne Rich, she adds that it is an attitude essential to "world-protection, world-preservation, world-repair."[2] The question posed by this Reclusian image is very much in this spirit: whether we can fulfill our historical destiny as an integral part of nature, awakened to the earth, allowing it to reveal itself to us, and playing our role in holding and caring for it.

It has been almost a decade since the first edition of this work appeared. Since then, the crucial significance of Reclus' vision of humanity as the developing self-consciousness of the earth has become increasingly clear, as the costs of the continued operation of the system of economic, political, and technological domination become more and more evident. The magnitude of these costs is most striking in regard to global ecological crisis. Researchers at the Stockholm Resilience Centre have formulated a conception of "planetary boundaries" within which human activity could continue without precipitating global ecological collapse, and concluded that "transgressing one or more planetary boundaries may be deleterious or even catastrophic due to the risk of crossing thresholds that will trigger non-linear, abrupt environmental change within continental- to planetary-scale systems."[3] They have defined nine such boundaries: climate change, ocean acidification, stratospheric ozone depletion, biogeochemical nitrogen and phosphorus cycles, global freshwater use, rate of biodiversity loss, land-system change, chemical pollution, and atmospheric aerosol loading. They suggest that three boundaries have already been passed and that most others are rapidly being approached.[4] Knowledge of such threats has expanded greatly over the past decade. This is exemplified by the successive United Nations Climate Change Conferences that focus the world's attention on global climate crisis, while the negotiations fail ever more miserably.

Global social crisis has followed a similarly tragic trajectory. The consumptionist culture of nihilism and the productionist system of technological domination have continued to colonize all areas of global society, while the nationalist, ethnic identitarian, and religious fundamentalist reactions to these processes continue to accelerate. Integral to these developments (both those internal to the dominant system and those generated in reaction to it) is the continual expansion by capitalism and the state of mechanisms of surveillance, control and annihilation. In view of the inability of the dominant system to significantly reform, much less radically transform itself in the face of global social and ecological crisis, Reclus' call for a many-sided social ecological revolution to replace the system of domination with engaged and compassionate communities in solidarity with humanity and nature seems increasingly prophetic.

The year after this book originally appeared, 2005, marked the 100th anniversary of the death of Elisée Reclus and the 175th anniversary of his birth. It also signaled the beginning of a surge of interest in his thought. A number of international conferences brought together researchers and activists to discuss Reclus' work and its relevance today. These included conferences on "The Geographer, the City and the World" at the University of Montpellier, France (dedicated to Reclus and French geographer Paul Vidal de la Blache); "Elisée Reclus and Our Geographies: Texts and Pretexts" at the University of Lyon, France; "Elisée Reclus, Nature and Education" at the University of Milan-Bicocca, Italy; and "Humanity and the Earth / L'Homme et la Terre: The Legacy of Elisée Reclus" at Loyola University in New Orleans. Such international events have continued to take place as interest in Reclus expands. The most important of these, "Elisée Reclus and the Geography of the New World," took place in 2011 at the University of São Paulo, Brazil, which has become a global center of Reclus research.

In recent years, the rate of publication of works on or by Reclus has grown exponentially. In the wake of the centennial, two new books on Reclus and two collections of Reclus conference proceedings appeared.[5] In the succeeding years, a new biography, a work on Reclus and colonialism, another on Reclus and the United States, and a brief introductory text have appeared.[6] In addition, anarchist geographer Philippe Pelletier recently published a massive volume on Reclus, Kropotkin, and Metchnikoff to follow up his earlier book on Reclus and anarchy.[7] In addition to the new French publications, Federico Ferretti, who is responsible for some of the most important research and writing on Reclus, has published two works in Italian.[8] New editions of Reclus' own works have included two collections of his writings on the Americas, a volume of his letters from prison and exile, a reprint of his "Great Globe" project, and a new edition of his anarchist political writings.[9] While translation of Reclus' writings into major languages has proceeded slowly, twelve volumes of Reclus' works have recently been published in Portuguese, and a collection in Spanish is forthcoming.[10] A documentary film on Reclus has also appeared recently.[11] In view of the rapidly growing interest in Reclus, this is an auspicious time for the new edition of the present work to appear.

We would like to reiterate our gratitude to the many who contributed to the project, and who were mentioned in the preface to the first edition. Prof. Ronald Creagh deserves further thanks for his contributions to this revised edition. In addition, we would like to thank Ramsey Kanaan and Craig O'Hara of PM Press for their encouragement and assistance, and John Yates for his skill and imagination in designing the cover of this edition. We are extremely grateful to PM Press for making it possible for this work to become readily available to its intended audience for the first time. With its publication of the first accessible edition of this work and the first comprehensive collection of writings of Gustav Landauer in English, PM Press has been instrumental in bringing to an English-speaking audience the work of the two greatest classical communitarian anarchist philosophers. Paraphrasing Hegel's famous statement about philosophy, "the Owl of Minerva takes flight at dusk," we might say that for anarchist philosophy today the Owl of Minerva increasingly takes flight at PM.

Preface to the First Edition

Elisée Reclus' life and ideas have been an inspiration to both of us ever since we first discovered his fascinating account of his voyage to New Orleans. We both have a strong interest in French culture and ideas and in the history of the French in America—an interest that was influenced by our Louisiana French family backgrounds. One of us has long been interested in anarchist theory and social ecology, and has written several books on these subjects. For these reasons, we were intrigued by this French anarchist geographer and his acute observations on the land of our ancestors, *la Louisiane*. We went on to translate the text of Reclus' voyage, which was published as *A Voyage to New Orleans: Anarchist Impressions of the Old South*.[1]

As we continued to study Reclus over most of the past decade, we found his writings not only interesting historically but also pertinent to today's world. We have both been active in the Green movement and in various ecological projects, and for a number of years we edited a magazine concerned with (among other things) bioregional culture and ecological politics. We were struck by the degree to which Reclus' ideas concerning the relationship between humanity and the earth, his view of history and the struggle for liberation, and his critique of various forms of oppression and domination were relevant to the theory and practice of political ecology. Reclus' efforts to put his inspiring ideals into practice in his personal life also impressed us greatly. We concluded that despite serious limitations in some areas, he has an important message of freedom, human love and solidarity, and reconciliation with nature that is as meaningful today as ever before. In fact, it takes on even more significance in an increasingly cynical age that is sorely in need of a vision of hope, social creativity, and the universal good. This book is the result of our desire to share Reclus' vision with others.

The project of selecting excerpts from Reclus' voluminous writings was rather daunting. His two most important works alone run to twenty-five volumes and more than twenty thousand pages, and he also wrote other important multivolume works. Furthermore, he contributed many articles to geographical journals, intellectual reviews, and popular political magazines, in addition to writing a number of widely circulated political pamphlets. The translated selections and introductory essays draw on many of these works, but most particularly Reclus' magnum opus of social geography and social theory, *Man and the Earth.* This 3,500-page book, which was the culmination of his life's work, has been almost unknown to the English-speaking world. Our translation makes key sections of this work available to English-language readers for the first time. We are also presenting the first English translation of a large section of Reclus' only full-length political work, *Evolution, Revolution and the Anarchist Ideal,* and several important short works written over a period of a half-century. In addition, the introductory essays offer the first extensive analysis of Reclus' social thought ever to appear in English. Our goal is to offer the reader a brief but comprehensive view of Reclus' life and work, and an appreciation of his importance to philosophy, social theory, and political thought.

The translation of diverse works published between 1866 and 1905 presented certain difficulties. The most demanding of our challenges was to produce a translation that would be readable for a contemporary audience but still capture the flavor of Reclus' nineteenth-century prose. His final work, *Man and the Earth,* presented the most formidable difficulties. According to his nephew Paul, Reclus completed the manuscript in 1903, and they worked together on editorial revisions of this vast six-volume work "between October 1903 and Reclus' death in July 1905."[2] Certain important discussions, while quite coherent and sometimes reaching the level of eloquence, never received the thorough editing they deserved. We have attempted to remain faithful to Reclus' meaning while achieving as much clarity as the texts allow.

We have consistently employed English cognates in certain cases in which Reclus' French usage strongly reflects his culture and historical epoch. We felt that it was important to use the generic "man," not only because it was the contemporary English equivalent of Reclus' "*l'homme*" but also because it expresses very well the tension, and indeed the conflict, between his anarchistic, liberatory aspirations and the conventional, and even conservative, conceptual framework he inherited. The same point applies to such terminology as the "conquest" of various goals as opposed to their "achievement"; the "discovery" of regions by Europeans rather than their "exploration"; and the description of certain cultures as "primitive," "savage," and "barbarian" rather than "tribal," "hunting and gathering," or "planting." At times, Reclus explicitly recognized the problematical nature of some of these terms, but he continued to use them, and they certainly express the classic modernist political sensibility that constitutes an important dimension of his outlook.

Language typical of the classical workers' movement and nineteenth-century radicalism has also been retained. For example, the term *maître* has usually been translated as "master," a term that frequently appeared in English-language anarchist prose of his time, in preference to "ruler" or other more contemporary terms. Reclus uses two terms, *camarade* and *compagnon*, for his fellow members of the revolutionary movement. We have translated these terms as "comrade" and "companion," for even though only the former is common in such a context in English, various cognates of the latter term have been very widely used in the international anarchist movement. Reclus' *pain* has consistently been translated as "bread," even when the terms "food" or "necessities of life" would seem more natural today. It was Reclus who gave Kropotkin the title for his famous work *The Conquest of Bread*. Although this phrase may now strike one as rather strange, it reflects very well the social imagination of European revolutionaries of the nineteenth century. Finally, it will be noted that in a few cases we have included the original French in brackets when the word is unusual, or when the original might usefully convey certain connotations to readers with some knowledge of French.

We would like to express our deepest thanks to our close friend and colleague Prof. Ronald Creagh of the Université Paul Valéry in Montpellier, France, for many hours of discussion of numerous points of translation, for his highly perceptive comments on the introductory essays, and for his assistance in locating important texts. We also wish to express our deep gratitude to M. Pierre Bravo-Gala for his generosity and enthusiasm in discussing our translation, for his friendship and hospitality, and for his many astute suggestions on the interpretation of some of Reclus' most perplexing passages. We would like to thank Prof. Gary Dunbar for his very helpful comments on the text and for his generous gift of invaluable research materials. We are grateful to Prof. Kent Mathewson for his encouragement and support, and to Mr. Pavlos Stavropoulos for encouraging us to expand our project to its present scope. We would also like to recognize Ms. Deborah F. Justice for outstanding editorial work on the text.

Many others contributed to our work through suggestions, assistance in locating materials, advice on technical issues in their fields, and general encouragement of our project. Included in one or more of these categories are Mr. Alvaro Alcazar of Loyola University; Prof. Myrna Breitbart of

Hampshire College; Prof. Maurice Brungardt of Loyola University; Prof. Bernard Cook of Loyola University; Mr. David Crawford; Mme. Françoise Creagh; Prof. Anatol Dolgoff; Ms. Patricia Doran, Loyola University Library interlibrary loan coordinator; Prof. Marie Fleming of the University of Western Ontario; Mr. Jeffrey Harrington; Mr. Tetsushi Hiruma; Ms. Riki Matthews; the late Rev. Thomas Mulcrone, S.J., of Loyola University; Prof. Tom Starnes; and Mme. Stéphane Tiné of the staff of the French Senate. John Clark would also like to thank the Loyola University Grants and Leaves Committee for travel assistance for research. He is also very grateful to Camille Martin for applying her considerable editorial skills to the chapters he wrote.

John Clark and Camille Martin
New Orleans, Louisiana, and Bayou LaTerre, Mississippi

PART I

An Introduction to Reclus' Social Thought

1

The Earth Story, the Human Story

Elisée Reclus begins his magnum opus of social theory, *Man and the Earth*, with the words "L'Homme est la nature prenant conscience d'elle-même"—"Humanity is nature becoming self-conscious."[1] Above this statement is an image of the earth, held in two upward-stretching hands. In an important sense, the purpose of that work, and indeed, of Reclus' entire life's work, is to draw out the implications of these words and this image. He wishes to trace the course of human history, showing the unity of development underlying the diversity of cultures and epochs, and then to situate the history of our species within the larger history of the planet. In doing so, he hopes to contribute significantly to the very process of the development of self-consciousness that he describes.

Reclus wishes in this way to help humanity discover its meaning as a historical being and as an aspect of the earth's larger processes of self-realization. It is his further hope that the discovery of these truths about ourselves can also help us to act consciously and responsibly as part of a developing human community and a developing earth community. In short, Reclus retells the story of humanity in the context of the story of the earth. He thus places his work firmly in the tradition of the great historical narratives.

This is a tradition that might seem outmoded today. The revolutionary "grand narratives" of socialism and communism have been widely discredited, and even the dominant "grand narratives" of capitalism, technological progress, and nationalism appear in an increasingly demythologized form. The power of brute facts (or certain social conditions that are ideologically mystified as "brute facts") seemingly banishes the great myths of progress and social transformation. Ironically, mystification displaces mythology.

We (or at least the "we" of the West and its dependencies) seem now to be living through a period between narratives, between myths, if indeed we have not reached the end of the history of myth. It is a time of nihilism, in which the quest for being and meaning is replaced by the struggle for power. Without a Golden Age to emulate or a utopia to create, we find ourselves seemingly trapped in a rather uninspiring if overawing present. Banality is raised to the level of the sublime. We retain bits and pieces of the fragmented myths of the past and increasingly find ourselves left with disconnected bits and pieces of self. At worst, we merely accumulate and discard; at best, we recycle.

Reclus lived in a strikingly different age, in the heyday of the Myth of Progress. Partisans of the system of domination exuded optimism, if not smug complacency. Its opponents bristled with righteous indignation and glowed with idealistic hope for revolutionary change. Today, such hope has largely been replaced by the spirit of resignation. Conservatism declines into hardened cynicism, while radicalism is reduced to resentful marginalization. Between the two reigns confusion.

In an age of resignation, any narrative of universal self-realization seems suspect—as evidenced by recent postmodernist critiques of the very idea of the "grand narrative." Critics allege that any attempt to discover a transhistorical universality, or even any unifying thread running through the fabric of history, betrays an intellectual will to power, cultural imperialism, or a disguised apology for the forces of domination. It is indeed true that such narratives are usually guilty of one or more of these charges. And granted, it is the function of critical thought to subject all interpretation to the most ruthless questioning.

Yet Reclus' anarchist "grand narrative"—for the very reason that it is self-consciously anarchistic and aims at the destruction of the system of domination—presents a challenge to such sweeping dismissals of the quest for a more comprehensive, holistic view of history. Although it is true (and not very surprising) that Reclus does not entirely escape the biases that plague the creators of universal narratives, there are fundamental differences between his project and almost all the others. One of these concerns the identity of the historical protagonist (the "subject" or agent of history). Reclus' universal subject is not, as one finds in the stereotypical "grand narrative," Western or "civilized" humanity engaged in a process of triumphant world domination. Rather, it is a global humanity, embedded in nature, yet undertaking an open-ended and creative project of liberatory self-realization.

Thus, Reclus can be looked upon as an early prophet of globalization. His significance today comes in large part from his presentation of an egalitarian, libertarian vision of globalization—a globalization "from below"—that offers a theoretical alternative to the dominant corporate and statist versions that now prevail. Writing in the 1870s, he foresees a future in which "equality will obtain in the end, not only between America and Europe, but also between these two and the other quarters of the world." In place of a world divided into a powerful, wealthy, and hegemonic core and a weak, poor, and dominated periphery, the world will have "its center everywhere, its periphery nowhere."[2]

Reclus' project in fact points beyond even the globalization of humanity, for he understood the globe as the whole earth, of which we are an integral part. We develop within and in relation to that whole in all its complexity. He takes a dialectical approach in which every phenomenon, including the phenomenon of humanity, is inseparable from other phenomena (geographical features of the land, other living beings, natural regions) to which it is related. An understanding of the world thus requires a simultaneous understanding of all the interconnected and interpenetrating factors. For as Reclus states, though we must always seek to understand the significance of each determining factor, "it is only through an act of pure abstraction that one can contrive to present a particular aspect of the environment as if it had a distinct existence, and strive to isolate it from all the others, in order to study its essential influence."[3]

Thus, from his dialectical perspective the unity of history must be discovered through an understanding of the diversity of phenomena, both natural and social. Accordingly, his account of the human story recognizes the integrity and specificity of the other—whether this other be a cultural or natural one. He recognizes various past cultures and many existing non-Western ones for their unique and enduring contributions to progress, and he rejects the reduction of these cultures to mere obsolete stages of development toward the higher social, political, economic, and intellectual achievements of the modern West. Nor does he depict the natural world as a mere backdrop for human history. Rather, nature is for him always an active presence, both encompassing humanity and remaining in intimate dialectical interaction with humanity throughout history.

Furthermore, Reclus' account of human history and earth history avoids the imposition of closure, and it always retains a moment of creativity, novelty, and openness. In the preface to *Man and the Earth*, he summarizes what he sees as "the three orders of facts that are revealed to us through the study of social geography, and which remain so constant amidst the chaos of things that one might well label them 'laws.'" These are "the class struggle, the quest for equilibrium and the sovereign decision of the individual."[4] In each case, the sweeping historical "law" is far from being a narrowly deterministic one. The class struggle for Reclus is a realm of creative, self-expressive activity on the part of the dominated and oppressed, not a result of their mere reactivity to social conditions. Moreover, the quest for equilibrium is a creative project of humanity in which human beings have over the course of history invented diverse modes of cooperation, mutual aid, and cultural self-expression. For Reclus, social disequilibrium arises from the lack of freedom and from the attempt to impose a static order on a dynamic social milieu.[5] He agrees with Proudhon that freedom is "the mother, not the daughter, of order," and he adds that domination necessarily engenders disorder. Finally, despite the strongly communitarian dimension of Reclus' anarchism, it is "the human person, the primary element of society," that is the source of "the creative will that constructs and reconstructs the world."[6] It is his hope that this creative freedom will lead humanity to a future society based on free association, which will synthesize social harmony and equilibrium with social diversity and spontaneity.

Reclus was a pioneer in the project of writing the story of the earth and of humanity.[7] His anarchist grand narrative is a sweeping account of the planet from its beginnings, through the course of evolution over the ages, and finally through the manifold episodes of the human story within the story. It depicts the intersection of natural history and human history—or as he terms it, the story of "nature becoming self-conscious." Integral to this history is an account of the forces of domination that emerge in human history, only to restrict the future self-realization of both humanity and nature. So needless to say, another central theme of his story is the long quest to overcome these forces of domination.

In exploring such themes, Reclus anticipates later critiques of the domination of humanity and nature, developed from the Frankfurt School through contemporary poststructuralist and radical ecological thought. We are now entering an era in which concepts of global social and ecological crisis become more familiar; ideas that were once limited primarily to the arcane realm of abstract social theory begin to pervade the larger culture. Consequently, the world-historical narrative that Reclus recounts may have even greater resonance today than it did in his own time.

As will be shown in the discussion that follows, Reclus' emancipatory vision of history is a sweeping one with universalistic dimensions, but it encompasses a social and ecological ethic that is based on a concern for the self-realization of all beings in their uniqueness and particularity, and a practice of love and care for those beings.[8] This ethic is perhaps summarized best (at least in regard to humanity) in a letter that Reclus wrote near the end of his life to a Protestant minister in Orthez, his hometown. He asserts that his own ethic embodies the highest ideals of Christianity, ideals that had, perhaps ironically, been betrayed by institutionalized religion but carried on by anarchism:

> It seems to me that as a libertarian socialist or, to be more precise, a communist anarchist, I am in many ways close to the Christian of the

> Gospels. Thus, I must neither call anyone "master" nor call myself "master" of anyone else. I must seek to live in a condition of equality with everyone, Jew or Greek, owner or slave, millionaire or beggar, without accepting any kind of supposed superiority or inferiority. I must adopt the old Christian maxim not to do anything to others that I would not want done to myself, and to do to others only what I would wish them to do to me. If I claim the right of personal or collective self-defense, nevertheless I will forbid myself any idea of vengeance, which is a primitive practice, and no hatred will arise in my heart, since it would be aimed at unfortunates who are already victims of atavism or a bad environment. Finally, and again like the Christian who is faithful to his name, I would love first of all the brother whom I see "before cherishing or adoring the unknown beings that I do not see."[9]

Reclus' life work was to prepare the way for a world in which all forms of domination—all "mastery"—would be abolished, so that humanity could live in a free community of equals founded on such a practice of active, engaged love and compassion.

2

The Anarchist Geographer

Elisée Reclus was born on March 15, 1830, in Sainte-Foy-la-Grande, a small town on the Dordogne River in southwestern France. His father, Jacques Reclus, was a minister in Sainte-Foy and a professor at the nearby Protestant college. He was, in effect, a Protestant among Protestants, deciding to leave the French Reformed Church to become the pastor of a "Free Church" in the town of Orthez. By leaving an established church, Jacques Reclus rejected, for the sake of his beliefs, the possibilities of personal advancement and greater material security for himself and his large family. According to Elisée's nephew and biographer, Paul Reclus, Jacques powerfully influenced his children by his dedication to his principles, by "putting communism into practice" in his daily life, and by showing himself, through his independence from official religion, to be a "precursor of anarchism."[1] Elisée echoes these sentiments when he says that his father "was not an ordinary man who is content to live in accord with the world: he had the strange fantasy of wishing to live according to his conscience."[2] Elsewhere, he notes that while Jacques at first dominated the Reclus children through his powerful personality, his lasting influence took the form of creating in them "the ideal of the unyielding conscience."[3] Elisée's own independence of thought and his quest for a just community were thus conditioned by his paternal heritage of religious dissent. Indeed, his anarchism can be seen as the ultimate Protestant revolt against two of the most dominant religions of the modern age: the deification of capital and the worship of the state. Moreover, his later ideal of a universal community of love is clearly a transformation of the concept of Christian love that he encountered in his early life.

Other familial influences on Elisée were also very important. His sense of dedication to the general good was fostered by the example of his mother, Marguerite Trigant, who inspired admiration within the family and the community for her tireless efforts in conducting a school for girls, while conscientiously raising thirteen children, eleven of whom reached adulthood. Marguerite also influenced her children through her knowledge of literature, her encouragement of good writing,[4] and her "deep love" for the family.[5] While Reclus later broke with his parents over their conservative religious views, both left enduring marks on his character and ideals. Moreover, his ties with the rest of the family remained unusually strong throughout his life. This was true above all in the case of his older brother, Elie, with whom he maintained a deep personal, political, and intellectual relationship throughout the course of their long lives. While Reclus later launched a fierce attack on the patriarchal authoritarian family, the family as a loving community of mutual aid and solidarity had a strong influence on his vision of the good society.

Reclus was educated primarily in Protestant institutions. At the age of twelve he was sent to the Moravian School in Neuwied, Germany, where he learned German, Latin, English, and Dutch. His budding cosmopolitanism was encouraged not only by his exposure to another culture and to diverse languages but also by his personal experience of nationalistic animosity and prejudice against foreigners on the part of many of his fellow students. These experiences contributed to his growing concern for universal justice and belief in human solidarity. He returned to France, attended the Protestant College of Sainte-Foy, from which he received the baccalaureate, and then went to the Protestant University in Montauban. By this time, the seventeen-year-old Reclus had already developed an interest in radical political ideas and was becoming increasingly disillusioned with his conservative Calvinist environment.

Looking back on this period, he remarks that he, his brother Elie, and their schoolmates began to broaden their horizons as they heard news from Paris "of political struggles" and "then, all at once, of the Revolution itself."[6] The growing rebelliousness of the Reclus brothers is evidenced by the fact that in the next year they were both expelled from the university for leaving school without permission to travel to the Mediterranean. For Elisée, this event perhaps signaled both his growing rejection of established institutions and his budding passion for exploring the larger world. Elie later described Elisée's reaction to his first view of the great sea as ecstatic. Despite his restlessness, Reclus managed to return to the school at Neuwied, where he taught briefly. He then completed his formal education with six months of study at the University of Berlin. This stay, though relatively brief, was crucial in his development, for it was in Berlin that he attended lectures by the famous geographer Carl Ritter, who greatly stimulated his interest in his future field of specialization.

Already during his student years, Reclus' political ideas were quite radical. In a manuscript dating from this period, the twenty-one-year-old expresses views that already quite clearly defined the course of his future anarchism and its underlying basis. He judges the goal of history to be "complete and absolute liberty," adding that such liberty will amount to nothing more than "colossal egoism" if it is not united with love.[7] "For each individual man," he asserts, "liberty is an end," but in a larger sense "it is only a means toward love and universal brotherhood."[8] Reclus' lifelong concern with a synthesis of the ideals of freedom and solidarity are thus already quite evident. Even at this early date the implications of his views were clear enough for him to state, in terms reminiscent of Proudhon, that "our destiny is to reach that state of ideal perfection in which nations will no longer need to be under the tutelage of a government or of another nation; it is the absence of government, it is anarchy, the highest expression of order."[9]

Reclus' conception of freedom had by this time already extended beyond the political into other realms, including the economic. He asserts that "political liberty is nothing without other liberties" and that freedom is meaningless "for those who despite their sweat cannot buy bread for their families, and for those workers who only incur new sufferings through the revolutions they themselves make."[10] He also anticipates his later critique of authoritarian socialism in noting that "some communist varieties [of socialism], in reaction against the present-day society, seem to believe that men ought to dissolve themselves into the mass and become nothing more than the innumerable arms of an octopus" or "drops of water lost in the sea."[11] Reclus holds, to the contrary, that community and solidarity can never be separated from liberty and individuality. In this his ideas are reminiscent of those of William Godwin, his great predecessor in the tradition of philosophical anarchism. Godwin also emerged from the tradition of Protestant dissent and like Reclus was heir to a legacy of deep concern for the inviolability of individual conscience and respect for personal autonomy.[12]

After leaving Berlin, Elisée joined Elie in a walk across France, from Strasbourg on the Rhine in the northeast, to Orthez in the extreme southwest corner of the hexagon. By this time, both brothers had developed a passion not only for advanced political ideas but also for radical political action. They were enraged by Louis Napoléon's coup d'état of December 2, 1851, and met to plan a march to the *mairie* of Orthez to organize resistance. Only a small group of would-be insurrectionists actually set out the next morning for the *mairie*, and even these few, one by one, abandoned their plan. By the time the dwindling revolutionary mob reached its destination, it consisted of only two participants, Elisée and Elie.[13] Although their rebellion had turned into a fiasco, the authorities seemed to take the matter seriously, and the Reclus brothers found it necessary to leave

France and take refuge in England. For Elisée, this flight began over five years of foreign travel, and it profoundly affected his future vocation as a geographer.

Reclus spent most of a year in England and Ireland, working first as a tutor in London and then as a farm worker near Dublin. During this period, he developed the idea of exploring the Americas with the intention of eventually establishing an agricultural community in cooperation with Elie and other friends. By early 1853, he had crossed the Atlantic and was living in Louisiana.[14] His *Voyage to New Orleans* recounts his passage through the Antilles, his trip up the Mississippi Delta, and his striking impressions of the city of New Orleans. It also chronicles an important stage in the development of his social and political ideas. After working briefly as a dockworker in New Orleans, he found a position as a tutor for the children of the Fortier family of Félicité Plantation. He spent most of his two and one-half years in Louisiana on this plantation, fifty miles up-river from New Orleans on the west bank of the Mississippi. His experience with the much-romanticized plantation society of the Old South produced in him above all a visceral reaction to the cruel inhumanity of slavery.[15] His revulsion toward the slave system was largely responsible for his decision to leave Louisiana. He wrote that he could not continue to earn money by tutoring the children of slaveholders and thus "steal from the Negroes who by their sweat and blood have earned the money that I put in my pocket."[16] His strong sense of personal moral responsibility is evident in his judgment concerning his relationship to the system of slavery. He concluded that by remaining in the plantation house even in the seemingly innocuous role of tutor and participating in such an institution, "it is indeed I who hold the whip."[17]

In addition to intensifying his hatred of racism, Reclus' visit to Louisiana also strengthened his belief in the inhumanity of capitalism. While his experiences in Europe had already led him to abhor the evils of economic inequality and exploitation, he discovered in America an economistic mentality that far surpassed anything he had experienced in more traditionalist European societies. He concluded that the spirit of commerce and material gain had deeply infected American culture and poisoned it. As he wrote to his brother Elie, he believed the country to be a "great auction house in which everything is for sale, the slaves and the owner into the bargain, votes and honor, the Bible and consciences. Everything goes to the highest bidder."[18] His loathing for the virtues of free enterprise continued throughout his lifetime.

After leaving Louisiana, Reclus spent eighteen months in New Granada (now Colombia), where he attempted unsuccessfully to realize his dream of a cooperative agricultural community. His efforts were doomed by yellow fever, inadequate planning, and his partnership with an elderly Frenchman who turned out to be untrustworthy. Reclus was reduced to penury by this disastrous undertaking and ended up "without the means even to buy a pair of shoes."[19] Despite the setbacks that Reclus experienced in the Americas, his travels on both continents contributed greatly to his development as a geographer. During his stay in Louisiana, he traveled up the Mississippi and into Canada, making observations that would be invaluable for his later writings on North America.[20] In addition, his visit to New Granada formed the basis for his book *Voyage to the Sierra Nevada de Santa Marta*.[21]

After six years of travel, Reclus decided to return to his home and family and to seek new opportunities in France. He returned with his idealism and creative energy seemingly unaffected by his adversities and with a wealth of experience that would be invaluable in his future vocation. His strong beliefs concerning the desirability of blending races and cultures were put into practice in his personal life when he married, in December 1858, Clarisse Brian, the daughter of a French father and a Senegalese mother. According to Paul Reclus, "there is not the slightest doubt that Elisée's stay in Louisiana formed in him the idea of marrying a daughter of the despised race."[22] To whatever degree this may have been his motive, the marriage was also based on personal affinity and was a happy one. Tragically, it ended after ten years with Clarisse's death, shortly after the birth of their third child, who also died soon thereafter. A year later, Reclus married an old friend, Fanny L'Herminez, according to anarchist principles—that is, without the intrusion of either church or state. This alliance proved to be his closest and most cherished relationship with any woman in his life since the two shared many common values, intellectual interests, and political commitments. There was a deep spiritual affinity between them comparable only to that which Reclus had with his brother Elie. Although Fanny died less than four years after the marriage, he was profoundly affected by her for the rest of his life and for many years included her name or initials as part of his signature. He later entered into another "free union" with his third wife, Ermance Beaumont-Trigant. This relationship was also a personally fulfilling one for Reclus, though it lacked the spiritual depth he had found in his relationship with Fanny.

The testimony of Reclus' friends and colleagues indicates that his egalitarian and cooperative ideas were practiced admirably in his personal life. His fundamental principles of solidarity and mutual aid were much more than political slogans. According to his friend and fellow anarchist geographer Peter Kropotkin, "the idea of dominating anyone at all seems never to have crossed his mind; he hated down to the smallest signs a dominating spirit."[23] This was true of his relationships not only with his wives but also with other members of his family and his broad circle of friends. He was widely praised for his great humility and his reluctance to present himself as a "leader" or "expert." While he became well known as both a scientist and a political writer and activist, he vehemently rejected the idea of having followers or of placing himself in a position of superiority. As he once wrote to a young woman who was a would-be disciple: "For shame. . . . Is it right for some to be subordinated to others? I do not call myself 'your disciple.'"[24] There are numerous stories of his interactions with others on terms of complete equality and of his unassuming participation in the more menial aspects of political work. Jean Grave notes that Reclus "was able to listen to objections from whatever source they came, and to answer them without any pride and without the sharp tone of one who issues decrees, and admits of no discussion."[25] His spirit of nondomination extended beyond human beings to all other creatures and, indeed, to nature as a whole. He could not bear the idea of mistreating any sentient being, and he practiced vegetarianism for most of his life, on ethical grounds.

During the 1860s, Reclus published many geographical essays in the *Revue des deux mondes* and other journals, and he completed the first of his three great geographical projects. *The Earth* is an extensive fifteen-hundred-page work in two volumes, published in 1868 and 1869.[26] This impressive study of the physical geography of the earth established Reclus rather early in his lifetime as an important figure in his field. In 1869, he published *The Story of a Brook*, a popular work that became a classic of nature writing for young people.[27] It was later followed by a companion work, *The Story of a Mountain*.[28] Another of Reclus' activities in this period was his work in the cooperative movement, largely in support of Elie's initiatives. The two brothers were responsible for the publication of the cooperativist journal *L'Association* and the creation of a mutual bank, La Société du Crédit au Travail. The journal's difficulty in finding a public and the collapse of the bank contributed to Reclus' increasing disillusionment with the cooperative movement.

For Reclus and his circle, the early 1870s were dominated by the events of the Paris Commune and its aftermath. Since he was over forty years of age at the time, he was exempt from military service during the Franco-Prussian War. Nevertheless, he volunteered for the National Guard, believing that it was necessary to defend the Republic against reactionary enemies. He served in the balloonist company of his friend Félix Nadar, but he did not see military action until after the Commune was declared. During the brief life of the revolutionary regime, he actively participated in both its politics and the defense of the city. As Paris fell, his column of the National Guard was captured by the Versailles troops. During the next eleven months, he spent time in fourteen different prisons and was tried and sentenced to deportation to New Caledonia.

Despite his refusal to submit to the new regime, and largely because of his prestige as a scientist and intellectual, his friends and supporters succeeded in having his sentence reduced to ten years in exile. As a result, he was allowed to immigrate to Switzerland. Ironically, this exile at the hands of a reactionary regime contributed powerfully to his development as a radical political theorist and a force within the European anarchist movement, for in Switzerland he began his association with the anarchists of the Jura Federation and developed close ties with the major anarchist theorists Bakunin and Kropotkin. After some initial differences in outlook, Bakunin and Reclus became close collaborators in the First International and in the anarchist movement (including Bakunin's International Brotherhood). Bakunin once said of the Reclus brothers that he had never known any persons more "modest, noble, disinterested, pure, and religiously devoted to their principles."[29] These principles were close enough to Bakunin's own that the three remained strong political allies until Bakunin's death. Elisée delivered a eulogy to the great revolutionary at his funeral in Berne in 1876.

It was also in Switzerland that Reclus began his greatest geographical work, the *New Universal Geography*.[30] This monumental achievement appeared in nineteen large volumes between 1876 and 1894. The reader is struck not only by the quality of the writing, which, according to Patrick Geddes, "raised anew geography into literature,"[31] but also by the expansive scope of this seventeen-thousand-page work, the exhaustiveness of its details, and the magnificence of its illustrations. Geographer Gary Dunbar, in his biography of Reclus, concludes that "for a generation" this work "was to serve as the ultimate geographical authority" and that it constituted "probably the greatest individual writing feat in the history of geography."[32] Reclus remained in Switzerland until 1890, heavily occupied with his extensive scholarship and political activities, and then finally returned to France after almost two decades of exile.

In 1894, he began a new phase of his career when he accepted an invitation to become a professor at the New University in Brussels. Reclus had originally been invited to teach at the Free University of Brussels, but because of increasing public reaction against anarchist "propaganda by the deed," he was judged too controversial, and the invitation was withdrawn. This chain of events produced considerable dissension within the Free University and contributed to the decision to found the New University.[33] Despite the rather dissident character of this institution, Reclus had some reservations about entering even the remotest corner of the academic world, having remained an independent scholar, following his own political and intellectual path, until quite late in life. He wrote that although the "noble war cry" of the New University was "Let Us Make Men!" he feared that "to a certain degree it would also contribute to making exploiters."[34] Despite these misgivings, he finally accepted the challenge with enthusiasm and was a great success, achieving renown as a teacher and winning the enduring admiration of many students.

At the conclusion of the *New Universal Geography*, Reclus comments that from "the myriad facts" of that vast work he would like to "extract a general idea, and thus, in a small volume written at leisure, justify the long series of books now ended without apparent conclusion."[35] The "small volume" turned out to be Reclus' final major work, the six-volume, thirty-five-hundred-page *Man and the Earth*.[36] This impressive undertaking constitutes a grand synthesis of Reclus' ideas on geography, history, philosophy, science, politics, religion, anthropology, and many other fields. While the work reinforced his reputation as a major figure in the history of geography, it also expanded social geography beyond the conventional limits of the geographical into a comprehensive worldview. Since his publisher had compelled Reclus to avoid in the *New Universal Geography* any lengthy "digressions" on social and political topics, he reserved many of his most important theoretical reflections for this final work.[37] It is thus both the culmination of his life's work as a social geographer and the most developed expression of his anarchist social philosophy.

Reclus was admirably consistent in integrating his libertarian and communitarian ideals into his personal life, his political activism, and his scholarly work. His enduring love of life, of other human beings, and of truth is expressed eloquently in a letter written March 25, 1905, only a few months before his death. At the age of seventy-five, though ill and growing increasingly weaker, he could still write of "two powerful attractions" that gave him the will to live. The first consisted of "affection, tenderness, the joy of loving, the happiness of having friends and of making them feel that one loves them, that one asks nothing of them but to let themselves be loved, and that every token of affection is a delight freely given." The second, he says, is "the study of history, the joy of seeing the interconnection of things. There is doubtless a strong element of imagination in this study, and deceptive Maya also leads us down many false paths. But it is another great joy to recognize one's errors."[38]

Reclus died in the countryside at Thourout near Brussels on July 4, 1905. It is reported that his last days were made particularly happy by news of the popular revolution in Russia. He expired shortly after hearing of the revolt of the sailors on the battleship *Potemkin*.

3

The Dialectic of Nature and Culture

It is likely that Reclus' most enduring intellectual legacy will be his contribution to the development of the modern ecological worldview and his role in the creation of radical ecological social thought.[1] More specifically, he is important for introducing a strongly ecological dimension into the tradition of anarchist and libertarian social theory. This tradition, like Western thought in general, has been marked by humanity's alienation from the natural world and its quest to dominate nature. Yet it has been, on the whole, more successful than most others in uncovering the roots of this alienation, looking beyond the project of planetary domination, and attempting to restore humanity to its rightful place within, rather than above, nature. Reclus made a powerful contribution to introducing this more ecological perspective into anarchist thought.

It is noteworthy that social geography had an impact on anarchist theory at the end of the nineteenth and the beginning of the twentieth century, just as social ecology has had a certain influence on anarchist thought in the late twentieth and early twenty-first century.[2] While this historical parallel is occasionally noted, the connection is usually made through reference to Kropotkin as a forerunner of ecological anarchism. Few commentators have understood that Reclus, much more than Kropotkin, introduced into anarchist theory themes that were later developed in social ecology and eco-anarchism. Indeed, Reclus explored these social ecological issues with considerable theoretical sophistication—more than a century ago.[3]

Béatrice Giblin, in her article "Reclus: An Ecologist ahead of His Time?" contends that Reclus "had a global ecological sensibility that died with him for almost a full half-century."[4] This sweeping generalization is in some ways even an understatement of the case. The kind of ecological perspective that Reclus developed, especially in his great culminating work, *Man and the Earth*, effectively disappeared from mainstream social thought early in the century and did not reemerge significantly until well into the 1970s, in response to growing public awareness of the ecological crisis. In the meantime, ecological thinking remained an undercurrent of anarchist and utopian thought and practice, as, for example, in the work of such communitarian groups as the School of Living of Ralph Borsodi and Mildred Loomis.[5] However, it did not become a central theme in anarchist and utopian theoretical discussion until the ideas of Paul Goodman and Murray Bookchin began to have a noticeable influence in the late 1960s.[6]

It has been noted that Reclus begins the first volume of his magnum opus of social theory with the epigraph "Man is nature becoming self-conscious."[7] This concept—literally, that humanity is "nature taking consciousness of itself"—captures the essence of Reclus' message: that humanity must come to understand its identity as the self-consciousness of the earth and that it must complete the process of developing this consciousness in history. In effect, he proposes a theoretical project of understanding more fully our place in nature and of unmasking the ideologies that distort it, and a corresponding ethical project of assuming, through a transformed social practice, the far-reaching moral responsibilities implied by that crucial position. On the basis of this approach, he seeks to explain the development of human society in dialectical interaction with the rest of the natural world, and he expounds a theory of social progress in which human self-realization and the flourishing of the planet as a whole can finally be reconciled with one another.

Reclus always had a strong sense of humanity's intimate connectedness to the natural world. Even in his early work, he eloquently describes humanity's character as an expression of the earth's creativity and our kinship with all of life. "We are," he says, "the children of the 'beneficent mother,' like the trees of the forest and the reeds of the rivers. She it is from whom we derive our substance; she nourishes us with her mother's milk, she furnishes air to our lungs, and, in fact, supplies us with that wherein we live and move and have our being."[8] Throughout his works, he remains true to this integral, holistic vision of humanity-in-nature. While his studies became increasingly scientific, technical, and minutely detailed, he never abandoned the aesthetic, poetic, and even spiritual aspects of his attitude toward nature but rather synthesized these dimensions in his far-ranging, integrative perspective. Such a fusion of forms of rationality and imagination that have so often been opposed to one another in Western thinking is one of the most noteworthy dimensions of Reclus' thought.[9]

Similarly, Reclus seeks to integrate a theoretical and scientific understanding of nature with an awareness of the practical implications of such an understanding. His social geography is a thoroughly political geography, constantly exploring the question of what one might call "the politics of self-conscious nature." Yves Lacoste, the contemporary French geographer who has perhaps done most to revive interest in Reclus, contends that while Reclus was "the greatest French geographer," he has been "completely misunderstood" because of the "central epistemological problem of academic geography: the exclusion of the political."[10] Lacoste finds it ironic that recent discussions of social geography systematically "forget" Reclus' massive six-volume work in which social geography is the "main thread."[11] The situation parallels in some ways the reception of social ecology and radical political ecology today. Such perspectives are sometimes granted validity to the extent that they point out that "all things are connected," including ecological and social realities, but they often lose credibility when they begin to explore the nature of that connection—and dare to find the roots of ecological crisis in the existence of the centralized nation-state and the corporate capitalist economy.

Such parallels should not be surprising, for the connections between Reclus' social geography and social ecology in particular are in many ways quite striking. To the extent that social ecology remains radically dialectical, one of its fundamental interpretive principles is the concept that every phenomenon incorporates within itself the history of that phenomenon. Reclus uses much the same concept to guide his social geography when he observes that "present-day society contains within itself all past societies."[12] He also applies it to human nature, expressing a variation on the idea that ontogeny recapitulates phylogeny. In his formulation, "man recollects in his structure everything that his ancestors lived through during the vast expanse of ages. He indeed epitomizes in himself all that preceded him in existence, just as, in his embryonic life, he presents successively various forms of organization that are more simple than his own."[13]

There is thus for Reclus a continuity of development in both natural and social phenomena, in which the earlier stages are preserved in the later ones. This does not, however, imply any sort of strict deterministic outlook. Rather, our knowledge of continuities and determinants is seen as contributing to the increased freedom that results from an accurate understanding of the nature of things. Interestingly, Reclus does not hesitate to recognize similarities between "monarchy" in human society and "monarchy" within animal species, as in the case of some primates species with groups having dominant individuals, or, as he depicts them, "recognized chiefs."[14] Bookchin, on the other hand, completely rejects any such attributions on the ground that many essential features of human hierarchies do not exist in the animal communities that are described as hierarchical. He seems to fear that the use of such terminology might imply that human institutions are biologically based and therefore not subject to social transformation. Reclus would certainly recognize the significant differences between human and primate hierarchies, yet he sees the use of such terminology as no threat to his anarchist principles or his hopes for humanity. In his analysis, such language draws attention to a certain continuity between phenomena in the human and natural worlds. Yet from the history of humanity one can learn that social hierarchies are contingent, historically developed institutions that may be rejected if human beings choose to organize their communities in other ways.

Although Reclus believes we can learn much about existing social phenomena through the study of the evolution of all forms of life, his primary focus is on discovering the nature of these phenomena through the examination of their evolution over the history of human society in particular. Such an analysis will guide us in understanding both the structure of and the contradictions within present-day societies. In his analysis of these societies, he discovers that each of them "is composed of superimposed classes, representing in this century all successive previous centuries with their corresponding intellectual and moral cultures," and that when they are "seen in close juxtaposition, their vastly differing conditions of life present a striking contrast."[15] Through the investigation of these classes, Reclus seeks to uncover certain fissures in the social structure that are usually concealed by the dominant integrative ideologies. It can thus be shown how the hidden legacy of social domination reveals itself in contemporary social conflicts.

For Reclus, it is necessary to develop a critical consciousness of past historical development if we are ever to transcend the legacy of domination. Such an awareness is a precondition for the conscious creation of a future collective history, a process conceptualized by Reclus as humanity's attempt "to realize itself through one form that encompasses all ages."[16] As the species comes to see itself as part of a historical and geographical whole, it attains both self-consciousness and a corresponding freedom. We gain the ability "to free ourselves from the strict line of development determined by the environment that we inhabit and by the specific lineage of our race. Before us lies the infinite network of parallel, diverging, and intersecting roads that other segments of humanity have followed."[17] It is thus by comprehending the great diversity of human experience that humanity can achieve a unifying vision of its own history.

Just as in society unity is achieved through a recognition of diversity, in nature a unifying harmony is attained through diverse and often discordant elements. Although the ecological perspective has often been identified with a rather one-sided emphasis on harmony, balance, and order, recent discussions in ecological theory have challenged the dominant (ecosystemic, "balance of nature") model. Indeed, some theorists, inspired by postmodernist thought, have embraced the opposite extreme, seeing only disorder and chaos in nature. Reclus long ago supported a more judicious and theoretically balanced dialectical view that avoids the extremes of overemphasizing either order or chaos.[18] There is indeed, according to Reclus, a harmony and balance in nature, but it is one that operates through a tendency toward discord and imbalance. He notes that "as plants or animals, including humans, leave their native habitat and intrude on another environment, the harmony of nature is temporarily disturbed"; however, these introduced types either die out or adapt to the new conditions, making a contribution to nature as they "add to the wonderful harmony of the earth, and of all that springs up and grows upon its surface."[19] Thus, to the extent that there is a "balance of nature" it is not a simple balance of elements but rather a complex balance of order and disorder.

Reclus' deeply holistic account of natural processes often prefigures contemporary ecological analyses. An example is his discussion of the function of forests in global ecological health. He laments the reckless and destructive actions of the "pioneers" of both North and South America, who burned huge expanses of ancient forest in order to establish agriculture, "at the same time burning the animals, blackening the sky with smoke, and casting to the wind ashes that scatter over hundreds of kilometers."[20] He notes that while this action was shortsighted even from a narrowly economic point of view, the great loss is that the forests have been prevented from playing "their part in the general hygiene of the earth and its species," which is "an essential role."[21] Reclus sounds strikingly contemporary in proposing a health model of ecological soundness in which human health is linked to the health of ecosystems. Using strongly organicist imagery, he suggests that the earth "ought to be cared for like a great body, in which the breathing carried out by means of the forests regulates itself according to a scientific method; it has its lungs which ought to be respected by humans, since their own hygiene depends on them."[22] He also incorporates the aesthetic dimension in his holistic view of nature when he describes the earth as "rhythm and beauty expressed in a harmonious whole."[23] His discussion is strongly reminiscent of Aldo Leopold's land ethic, which stresses concern for "the health of the land," in the sense of "the capacity of the land for self-renewal," and famously defines "rightness" in terms of what "tends to preserve the integrity, stability, and beauty of the biotic community."[24]

Recent ecological thought has devoted a great deal of attention to the problem of anthropocentrism, a view that places human beings in a hierarchical position over all other beings and reduces all value in "external nature" to a merely instrumental one in relation to human ends. Reclus may sometimes sound rather anthropocentric, particularly when he focuses on the various "conquests" involved in human progress. However, his social geography actually constitutes a great step in the direction of incorporating humanity fully into the life and history of the planet. What is striking about his viewpoint is the degree to which he was able to transcend many of the dominant ideas of his century in shifting from an entirely human-centered to a more earth-centered perspective. The comments of an early commentator, Edward Rothen, in a memorial tribute, indicate the extent to which Reclus' approach could seem even in his own day to constitute a break with the dominant human-centered ideology. According to Rothen, Reclus "completely rejected that anthropomorphism which made man, the image of God, the sovereign of a world created only for the satisfaction of his needs and his whims."[25] Reclus "thought it stupid to deny a soul to animals, to plants and to all that is still termed 'insensible matter,' as if such matter could be found anywhere in the universe."[26] His view constitutes, according to Rothen, an "infinite pantheism" that "perceives an immense solidarity between all that lives."[27] Considering his recognition of the continuity and underlying unity of all being, and the awe with which he contemplated nature, this certainly captures an important dimension of Reclus' outlook. Reclus did not choose to use the term "pantheism" to refer to his worldview. Nevertheless, one finds in his works such passages as the following, from his chapter on "Bathing" in his *History of a River*: "It seems that I have become part of the surrounding milieu; I feel as if I am one with the floating aquatic plants, one with the sand swept along the bottom, one with the current that sways my body."[28] The sensibility expressed here is certainly close to what is sometimes called "nature mysticism" or "pantheism."

What is clear is that Reclus wished to situate humanity within the context of a larger reality of which it is a part. Far from being anthropocentric, Reclus' view of humanity's place in nature is dialectical, critically holistic, and developmental. In a sense, it might be called an "emergence" theory, if it is understood that for him humanity is emerging *within* nature rather than *out of* it. His analysis in some ways parallels the division of the natural world made by Bookchin and various other social and political ecologists who adopt the ancient distinction between a "first nature" and a "second nature," corresponding more or less to the natural world and the social world, both of which are seen as developing dimensions of a larger "nature."[29]

Reclus also delineates such realms of being in nature, but his analysis is a bit more complex than these. There is, on the one hand, that sphere of nature that can exist independently of humanity, and that had, indeed, existed for eons before nature began to "become conscious of itself" through the development of humanity. As humanity emerges, it remains in intimate interrelationship with an external sphere of nature, and the complex relationships of interdependence between the two realms take on an increasingly planetary dimension. Reclus calls the realm of natural being that has arisen and related itself to the rest of nature "the human social milieu."

However, the social world does not constitute for Reclus merely a "second nature," for it is itself dual and thus might be said to encompass both a "second" and a "third nature." He calls the former "the static milieu" or "the natural conditions of life," while he labels the latter "the dynamic milieu" or "the artificial sphere of existence." The first sphere, even though it is shaped by human culture, constitutes our most immediate embeddedness in nature and thus has a certain degree of natural necessity. The second sphere is much more subject to human direction and is therefore more a realm of social contingency. For Reclus, there is "a quite marked distinction between the facts of nature, which are impossible to avoid, and those which belong to an artificial world, and which one can flee or perhaps even completely ignore. The soil, the climate, the type of labor and diet, relations of kinship and marriage, the mode of grouping together, these are the primordial facts that play a part in the history of each man, as well as of each animal. However, wages, ownership, commerce, and the limits of the state are secondary facts."[30] In defense of the contingent nature of the institutions he associates with "secondary facts," he observes that many earlier societies managed to exist without them. He argues for the theoretical priority of the "static milieu," since it has always existed and has often had a determining force in social affairs. Although he admits that "quite often in the case of individuals the artificial sphere of existence prevails over the natural conditions of life," he thinks that "it is necessary to study the static milieu first and then to inquire into the dynamic milieu."[31]

The subtle relationship between the two spheres is, however, dialectical. For Reclus, this means that the influence of nature and of the "static milieu" is much greater than historians and social theorists have recognized. In the development of society over history "nothing is lost," because "the ancient causes, however attenuated, still act in a secondary manner, and the researcher can discover them in the hidden currents of the contemporary movement of society."[32] While superimposed political and economic factors are often given primary recognition as social causes, "this second dynamic milieu, added to the primitive static milieu, constitutes a whole of influences within which it is difficult, and often even impossible, to determine the preponderance of forces. This is all the more true because the relative importance of primary and secondary forces, whether purely geographical or already historical, varies according to peoples and ages. Once again, a phenomenon—including even the social whole—can only be understood as the cumulative product of its entire history. Indeed, humanity itself, "with all its characteristics of stature, proportion, traits, and cerebral capacity," is "the product of previous milieux multiplying themselves to infinity" since the origins of the species.[34]

In short, we reflect the earth and the regions of the planet in which we arose and developed. In Reclus' words, "The history of the development of mankind has been written beforehand in sublime lettering on the plains, valleys, and coasts of our continents."[35] While bioregionalism has only recently reintroduced this concept and brought it to the center of ecological thought, Reclus long ago developed the theme that human beings are, at the most fundamental level, regional creatures.[36] The relationship between humanity and the earth and its regions is a dialectical one, resulting from mutual interaction, as the earth expresses itself through humanity, and as humanity acts upon the earth. For Reclus this interaction includes not only harmonious interaction but also humanity's struggle with the rest of the natural world. It is important to recognize that "the accordance which exists between the globe and its inhabitants" cannot be described adequately through a one-sided focus on terms like "harmony," "balance," and "oneness" that stress the existence of order since that very order "proceeds from conflict as much as from concord."[37] Once again we find in the world a dialectic between order and chaos, in this case in the human relationship to nature.

Reclus is especially interested in analyzing the side of the interrelationship between humanity and nature that has been neglected by much of social thought throughout the modern period: the conditioning of the "social" by the "natural." His position on this subject should not be confused with the tradition that begins with Montesquieu's famous reflections on the influence of climate on society.[38] In such discussions, the appeal to natural influences becomes little more than an attempt to give an "objective" basis to the writer's social and cultural prejudices, so that characteristics attributed to various peoples become essential qualities dictating strict limits for possible social change. This tradition culminates in such theories as Huntington's "human geography," in which the appeal to nature becomes the ideological justification for white supremacy and European hegemony.[39]

Reclus' analysis should be distinguished from such views not only on the basis of his differing value commitments but also by his radically different methodology. In stressing the dialectical relationship between nature and culture, he focuses on the interaction between many natural and social factors in shaping human society, on the inevitability of change and transformation, and on the open-ended character of human and natural history. Far from attributing inherent, immutable qualities to peoples and cultures, he hopes that by understanding the determinants of the social world, all peoples can ultimately transform themselves into active, conscious agents in shaping their own liberation and self-realization, and that of the entire planet. His radically libertarian analysis illustrates the fact that the investigation of the influence of the natural world on cultural practices and social institutions does not necessarily have reactionary, authoritarian, or racist implications.

An example of Reclus' analysis of the influence of natural geography on social institutions is his treatment of the history of ancient religions. He hypothesizes that the monotheism of the ancient Near East reflects the austere character of that region's terrain. He remarks that one might generalize "that throughout the Semitic countries the splendid uniformity of tranquil spaces, illuminated by a violent sunlight, must have contributed mightily to giving a noble and serious turn to the concepts of the inhabitants. They learned to see things simply, without searching for great complications."[40] He contrasts this unifying vision to the unity-in-diversity expressed in Indian religion and suggests that the latter corresponds to the natural features of India. Near Eastern mythology "bore no resemblance to the chaos of divine forces leaping out of nature in infinite variation that one finds in India, with its high mountains, great rivers, immense forests, and climate whipped into rages by the abundant rains and the fury of storms."[41] Reclus notes that the "Hindu spirit" also perceived an underlying order and unity but that it naturally expressed this "single force" in "an infinite variety" of manifestations.[42]

Reclus also discusses the dialectic between nature and humanity, geography and society, in his discussion of the development of early Greek society. He notes that Greece consists of many basins or watersheds (to use bioregional terminology) divided from one another by rocky or mountainous terrain, and that "the features of the ground thus favored the division of the Greek people into a multitude of independent republics."[43] On the other hand, the ubiquity of the sea, with the long coastline, many inlets, and surrounding islands, exerted a unifying force. The sea "acts as a binding element" and has "made the maritime inhabitants of Greece a nation of sailors—amphibiae, as Strabo called them."[44] There is thus a dialectic both between the unifying and diversifying natural factors, and also between these natural factors and the social ones.

Reclus does not, it should be stressed, attempt to reduce the complexity of social phenomena, whether in ancient societies or any others, to a mere reflection of geographical qualities. Indeed, he often puts great emphasis on the significance of the economic, the technical, and other "material" determinants, not to mention the political and cultural ones, in shaping all aspects of society. Reclus lived in an age in which social analysis tended toward either an idealism in which material determinants were ignored or a materialism in which economic and technological determinants were attributed almost exclusive importance. In this context (which encompassed not only mainstream ideological thought but also most of radical social theory, in both its Marxist and anarchist varieties), Reclus wished to rectify the general neglect of the influence of the natural world on human history. His thought is therefore noteworthy for the degree to which it draws attention to natural and geographical factors that have been of historical importance in shaping human societies and that still exercise an influence both immediately and through the sedimentation of their effects within inherited social institutions.

Nature thus shapes humanity at the same time that humanity reshapes the natural world. While modern civilization has devoted much attention to the latter side of this dialectic, the power of humanity to transform nature, it has failed, however, to recognize the moral significance of nature as a dynamic realm of meaning, value and creativity. With this moral failing in mind, Reclus therefore launches a scathing critique of humanity's abuse of the earth. In "The Feeling for Nature" he writes of the "secret harmony" that exists between the earth and humanity, warning that when "reckless societies allow themselves to meddle with that which creates the beauty of their domain, they always end up regretting it."[45] He warns that when humanity degrades the natural world, it degrades itself. His analysis of this phenomenon is reminiscent of the view of Thomas Berry, who argues that the diversity and complexity of the human mind reflects the richness and complexity of the earth and its regions, so that in damaging the earth we harm ourselves not only physically but in our "intellectual understanding, aesthetic expression, and spiritual development."[46] Reclus states that "where the land has been defaced, where all poetry has disappeared from the countryside, the imagination is extinguished, the mind becomes impoverished, and routine and servility seize the soul, inclining it toward torpor and death."[47] Of course, Reclus does not neglect the more obvious material damage to human society caused by ecological degradation. He notes that "the brutal violence with which most nations have treated the nourishing earth" has been "foremost among the causes that have vanquished so many successive civilizations."[48]

Reclus believes that despite such abuses of the past there are many ways in which humanity can pursue its own good while achieving an ecologically sound and ethically grounded relationship with the natural world. One means of achieving this goal is to grasp the close link between the ethical and the aesthetic and apply this understanding to our social practice. He believes that whether or not we carry out our ethical obligations to the natural world will have much to do with our aesthetic appreciation of it. An ugly, degraded world will not be fulfilling to human beings, while a beautiful one will contribute to our own satisfaction and self-realization. One of his most eloquent statements of the connections among the human good, human ethical choice, and the beauty of nature is found in his *History of a Mountain*. He says that

> every people gives, so to speak, new clothing to the surrounding nature. By means of its fields and roads, by its dwellings and every manner of construction, by the way it arranges the trees and the landscape in general, the populace expresses the character of its own ideals. If it really has a feeling for beauty, it will make nature more beautiful. If, on the other hand, the great mass of humanity should remain as it is today, crude, egoistic and inauthentic, it will continue to mark the face of the earth with its wretched traces. Thus will the poet's cry of desperation become a reality: "Where can I flee? Nature itself has become hideous."[49]

Nevertheless, human transformative activities need not have such negative effects. Doing what is right from an ethical perspective can be identical with the creation and preservation of the beauty and integrity of the natural world. We can, Reclus says, find beauty in "the intimate and deeply seated harmony that exists between our own work and that of nature."[50] Thus, as we contribute to the flourishing of the natural world we make our own lives richer and more fulfilling. It is quite possible to "assist the soil instead of inveterately forcing it," and to achieve "the beautification as well as the improvement of his domain," by giving "an additional grace and majesty to the scenery which is most charming."[51] When this is done, human creative self-expression will be in accord with the creative self-expression of nature. We will have succeeded in "mak[ing] our existence as beautiful as possible, and, as best we can, adapt[ing] it to the aesthetic conditions of our environment."[52]

Reclus gives a number of examples of the ways in which humanity cooperates with the earth in producing goodness and beauty, rather than ruthlessly seeking to impose its will upon it. Although agriculture always involves significant human transformation of the landscape, there is no need for it to be a process of mining the soil. Reclus believes that it is quite possible for farmers to "comprehend" the land and to "humor" it by discovering which crops suit it best, and he contends that some forms of traditional farming have achieved this goal. He praises the Shaker communities for their symbiotic, mutualistic practices that make agriculture a "ceremony of love" in which all aspects of nature are "cherished."[53] Writing in the 1860s, he remarks that there are also good examples in Europe of the way in which agricultural productivity can be reconciled with the beauty of the landscape. He remarks that "a complete alliance of the beautiful and the useful" has been attained in certain areas of England, Lombardy, and Switzerland, places where agriculture is in fact "most advanced."[54] As other instances of such a beneficial alliance, he cites the draining of marshes in Flanders to produce farmland, the irrigation of the barren Crau region, the planting of olive trees along the slopes of the Apennines and Alps, and the replacement of Irish peat bogs by diverse forests.[55] In *History of a Mountain* he notes that ancient alluvial deposits near the Pyrenees were transported by canal in order to build up and enrich the "naked plains" of the Landes, or moors, of southwest France. He comments that in such undertakings one "certainly" must see "considerable progress."[56]

Reclus makes a strong case that such undertakings have increased natural beauty of certain kinds in addition to being useful for human society. He can thus be looked upon as a forefather of bioregional agriculture as later developed in the work of thinkers such as Wendell Berry and Bernard Charbonneau.[57] However, it seems that in discussions such as those just cited, Reclus exhibits a bias in favor of more humanized rather than relatively wild landscapes. In general, he seems much less sensitive to the natural beauty of the austere terrain of rugged mountains and plains or the rich wildness of a swampland than to the appeal of more pastoral landscapes. Similar questions have been raised concerning later theories that tend toward an "ecological humanism," as, for example, Bookchin's version of social ecology. However, if social geography and social ecology seem at times to exhibit a one-sided pastoralism, this is a quality of specific versions of these theories, rather than a fundamental limitation of either. Both theories in their strongest formulations encompass a dialectical view of the relationship between humanity and nature and a grasp of the importance of nondomination, spontaneous development, and unity-in-diversity in the self-realization of the whole. They are therefore in principle fully capable of recognizing the importance of both humanized and pastoral landscapes as well as wilder and biologically more diverse ecosystems.[58]

Throughout Reclus' works, there is a tension between his expression of an emerging holistic, ecological perspective and his retention of certain aspects of the dualistic, human-centered outlook that was so common in his age. In an early work, he exhibits the latter tendency rather strongly when he remarks favorably that science is "gradually converting the globe into one great organism always at work for the benefit of mankind."[59] This rather extravagant conception of the earth's processes as a vast conspiracy to benefit our species is far from Reclus' later, more dialectical analysis.

There, humanity is integrated into the planetary whole as the consciousness of the earth, and the healthy functioning of the earth's metabolism is seen to benefit humanity as one part of that flourishing whole. The idea that science might control the entire earth in any way vastly exaggerates the power of technological processes. Reclus claims that these processes have the capacity to make the earth into "that pleasant garden which has been dreamed of by poets in all ages."[60] Such an image of the earth errs through an overemphasis on stasis, neglecting the element of dialectical tension that must always characterize human confrontation with the otherness of nature. This imagery tends to legitimate the destruction of the ecologically necessary wildness and freedom of the natural world, and to idealize a domesticated, highly humanized natural world that is far from being an authentically ecological conception. Fortunately, such tendencies became more muted in Reclus' later works, though they do not disappear entirely.[61]

It should be recognized that Reclus was from the outset a forceful critic of various assaults on nature that were accepted with complacency by his contemporaries. He judges that in civilization's dealings with nature, "everything has been mismanaged," so that what is left is "a pseudo-nature spoilt by a thousand details—ugly constructions, trees lopped and twisted, footpaths brutally cut through woods and forests."[62] He judges that in view of humanity's ravaging of the natural world, it will be necessary to undertake an extensive process of ecological restoration, a topic that has only recently gained widespread attention in ecological thought with the rise of conservation biology. Reclus states that "a reckless system has defaced [nature's] beauty," and it will therefore be necessary for "man" to "endeavor to restore it" and to "repair the injuries committed by his predecessors."[63]

Reclus sees the project of moving from an exploitative relationship to nature to an ecologically sound one as having both subjective, ideological as well as objective, institutional aspects. He points out that human interaction with nature has not been guided by "a sentiment of respect and feeling" for nature but rather by "purely industrial or mercantile interests."[64] For a fundamental change to take place in humanity's relationship to nature, a revolution in values must certainly take place. But the ideological transformation that will result in the triumph of "respect and feeling" can only succeed if there is a complementary process of social transformation, a change that would overturn the dominance of those "industrial or mercantile interests." According to Reclus, a "complete union of Man with Nature can only be effected by the destruction of the boundaries between castes as well as between peoples."[65] This implies for Reclus (as will be discussed in chapter 6) the destruction of the system of economic inequality and exploitation embodied in capitalism, the system of political domination inherent in the modern state, the system of sexual hierarchy rooted in the patriarchal family, and the system of ethnic oppression rooted in racial hierarchies. In short, the domination of nature will continue as long as humanity remains under the sway of a vast system of social domination.

In his analysis of the effects on nature of such an exploitative society, Reclus showed a level of awareness of the dangers posed by the destruction of biodiversity and by ecological disruption that was unusual for his time. In *The Earth*, he presents examples of the extinction of species caused by human "destruction," "slaughter," and "butchery," concluding that human activity has caused a "rupture in the harmony primitively existing in the flora of our globe."[66] Long before wilderness preservation became an organized movement with the establishment of the Wilderness Society in 1936, and indeed even before the establishment of the first national park in the United States in 1872, Reclus was in the 1860s warning of the dangers to ancient forest ecosystems in North America. He laments the destruction of "colossal" and "noble" trees such as the sequoias of the West Coast, a process that has resulted in "perhaps an irreparable loss" in view of the "hundreds and thousands of years" that will be necessary for the forest's regeneration.[67] In *History of a Mountain* he attacks the clear-cutting of forests and asks whether one "is not tempted to curse" those who carry out such logging practices.[68] He also discusses the damage produced through the introduction into ecosystems (whether by intention or negligence) of exotic plants and animals, without consideration of their effects on ecological interrelationships. Here again, he focuses on a major ecological problem that has only recently gained widespread attention among those concerned with environmental issues. Reclus cites the poignant comment by the Maori of New Zealand that "the white man's rat drives away our rat, his fly drives away our fly, his clover kills our ferns, and the white man will end by destroying the Maori."[69]

On the other hand, Reclus' discussions of demography and population growth seem to show less than adequate ecological insight, at least from today's perspective. It was his opinion that the human population of 1.5 billion in his time was not only easily supportable but even "still very minimal, relative to the habitable surface of the earth."[70] He did not seriously consider the possible impact on the biosphere if the human population were to double several times over the next century. At one point, he minimizes the significance of increases in human population by noting that if each person were given a square meter of space, everyone could fit into the area of greater London. Such a fact is, of course, entirely irrelevant from the standpoint of social geography. We could stand several persons in each square meter, some even on the shoulders of others, without learning very much about the interaction between human communities and the earth.[71]

Fortunately, his discussion of population is often much more nuanced than this, though still tinged with progressivist optimism. He is well aware of the fact that there is no optimal human population that can be calculated by means of arithmetic and plane geometry, or even discovered through more complex natural and social sciences. In this recognition, he was already far ahead of some contemporary advocates of simplistic conceptions of "carrying capacity."[72] He notes that if the world consisted of a population of hunters, the earth could perhaps support a population of only 500 million, or one-third the actual population at the time he was writing. He cites various estimates of the possible sustainable human population and comments favorably on Ravenstein's view that a population of six billion is a possible limit.[73] Nevertheless he expresses skepticism about all such estimates since there are numerous variables that cannot be predicted with any certainty. As an example he cites changes in methods of production, most notably in the area of agriculture. In his view, such changes would probably allow a much greater human population to be supported. He believes that when farming attains "the intensive character that science dictates," the population will increase at "a completely unforeseen rate" and that "the expanse of good land, which is presently quite limited, cannot fail to grow rapidly, whether through irrigation, drainage, or the mixing of soils."[74] Reclus did not consider the possibility that if vastly increased social and ecological costs of increased technological development led to a slowing of growth in productivity, the per capita supply of arable land dwindled with population growth, and ecological degradation caused the quality of the soil to deteriorate, exactly opposite conclusions concerning population would follow. Today, the significant slowing of population growth in much of the global South can be seen as a response to such problems, which have been aggravated by the additional burdens imposed by the neoliberal restructuring of the global economy.

Reclus shared with many of his progressivist contemporaries certain pronatalist attitudes and saw a decline in birth rates in parts of Europe as a sign of decadence. He moralizes about the fact that in the more affluent areas, natality drops drastically. He cites the examples of the *départements* of l'Eure and Lot-et-Garonne, where the death rate had surpassed the birth rate for most of the century, despite the fact that these are among the *départements* "whose soils have the greatest fertility."[75] He attributes to the egoism of affluence the failure of the citizens to reproduce at a level that he thinks appropriate, and he presents the phenomenon as an example of how under capitalism the pursuit of individual self-interest conflicts with the general good. He notes that proprietors who fear the division of their land among numerous heirs find that having few offspring better serves their self-interest, and that the same holds for functionaries with modest incomes who want to move up in social status.[76] No doubt there is truth in his observations. But what he fails to note is that where egoism reigns, all social phenomena (including both the desire for offspring and the desire to limit the number of offspring) take on an egoistic coloring, and that their egoistic character in such a context says little about these phenomena "in themselves."

Despite his pronatalist tendencies, Reclus did not share the widespread view that an increase in population was an unmixed blessing to society. He says that although "growth in numbers has been, without doubt, an element contributing to civilization, it has not been the principal one, and that in certain cases it can be an obstacle to the development of true progress in personal and collective well-being, as well as to mutual good will."[77] It is likely that he would see significant population growth much more as an "obstacle" today, as ecological devastation accelerates, as the accompanying social crisis intensifies, and as a rapidly increasing human population has now surpassed the limit of six billion that even he, living in an optimistic age, considered plausible. Moreover, the conditions of production have changed in a sense contrary to the one he hoped for: their development under conditions of capitalist hegemony shows little promise of bestowing abundance on a rapidly expanding global population, while it threatens to destroy the biotic preconditions for supporting existing human and nonhuman populations at an "optimal level," if indeed at any level at all.

An area in which Reclus was far in advance of his time, and in which he anticipated current debate in ecophilosophy and environmental ethics, is his effort to raise both ethical and ecological issues concerning our treatment of other species. His ideas are important in view of the fact that he was not only a pioneer in ecological philosophy but also an early advocate of the humane treatment of animals and of ethical vegetarianism. Even today, after several decades of discussion of "animal rights" and "ecological thinking," there are few theorists who have attempted to think through the interrelationship between the two concerns. Yet more than a century ago, Reclus offered some highly suggestive ideas about how a comprehensive holistic outlook might encompass a serious consideration of our moral responsibilities toward other species.

Reclus observes that all social authorities, in addition to public opinion in general, "work together to harden the character of the child" in relation to animals used for food.[78] This conditioning, he says, destroys our sense of kinship with a being that "loves as we do, feels as we do, and might also progress under our influence, if it does not regress along with us."[79] Much like the utilitarian defenders of animal welfare since Bentham, he objects to the suffering inflicted on individual animals raised for food, but from a larger perspective he also censures the injury caused to the species by the process of domestication. The flourishing and adaptive development of species that takes place in the wild is halted and then reversed, as the animal is increasingly reshaped or reengineered in conformity with its single role as a food resource.

As has been noted, Reclus links the ethical and the aesthetic in his analysis of our relation to the phenomena of nature. He observes that the abuse of animals that we find to be morally repugnant is also repellent to our sensibilities. In this connection, he touches on the question of intrinsic value, a concept that is central to current debates in environmental ethics. He states that the "regression" of animals under human influence "is indeed one of the most deplorable results of our carnivorous practices, for the animals sacrificed to man's appetite have been systematically and methodically made ugly, weakened, deformed, and degraded in intelligence and moral worth."[80] Such a reduction of "moral worth" might refer to two aspects of the moral problem: first, that human treatment of animals reduces them to a level at which their lives and experience seem less valuable to human beings; and second, that the "debasing" treatment to which they are subjected reduces the animals' capacity for the attainment of their own peculiar good and for experience that has value in itself. Reclus' general discussion of animal issues indicates that he has both aspects in mind.

The importance of ethical vegetarianism for Reclus is that it "recogniz[es] the bonds of affection and kindness that link man to animals" and "extend[s] to our brothers who have been dismissed as inferior the feelings that have already put an end to cannibalism within the human species."[81] In a letter he explains that there are spheres of moral obligation extending out from one's own society, to humanity as a whole, and finally to all sentient beings:

> For my part, I also include animals in my feeling of socialist solidarity. But I also say to myself: everything comes in degrees and our primary obligations begin immediately around us! Let us realize justice in the largest circle in which we can possibly do so: in the civilized circle first, then in the human circle. Every partial realization of an ideal increases our sensitivity and delicacy and makes us more capable of realizing a larger ideal. All that we accomplish for our neighbor moves us closer to those who are now distant from us. I am firmly confident that our harmonic society should embrace not only humans but also all beings that have consciousness of their lives.[82]

Though Reclus refers in such passages to an "extension" of our moral sentiments, his position goes far beyond the "moral extensionism" of certain ethical theorists who merely apply conventional, nonecological ethical concepts to nonhumans. Reclus instead undertakes a fundamental rethinking of the ethical. He believes that our treatment of other species reflects our level of awareness of our connectedness to the whole of nature and of our development of feelings that are in accord with such awareness. In his view, our growing knowledge of animals and their behavior "will help us to delve more deeply into the life sciences, increase our knowledge of the nature of things, and expand our love."[83] Elsewhere he expresses his "fervent love of the justice that extends to all that lives, to the entire expanse of beings,"[84] and he refers to "the bond of solidarity" that unites him not only with "those beings that I respect and love" but also with "all that lives and suffers."[85] For Reclus, we develop morally as the scope of our knowledge expands and as our attachment to the larger whole of life is strengthened.

Once again, the centrality of the concept of love to Reclus' worldview is evident. His view of human moral development is noteworthy in relation to recent discussions of the distinction between the ethics of abstract moral principles and the ethics of care.[86] Reclus is unusual among nineteenth-century radical social thinkers in that he focuses so strongly on the importance of the development of moral feeling, compassion, and the practice of love and solidarity in everyday life. In his time, much of the radical opposition to the dominant order was fueled by a sense of injustice and outrage at the oppression and exploitation produced by that system. While this opposition certainly had an authentic ethical dimension, it also succumbed to the reactive mentality and spirit of resentment that Nietzsche so perceptively diagnosed in many versions of socialism, communism, and anarchism. Reclus' outlook achieves a remarkable synthesis between, on the one hand, the concern for justice, knowledge and rationality, and on the other hand, the need for social solidarity and the development of care and compassion. In this, he has much in common with contemporary feminist ethicists who wish to restore the balance between these two sets of concerns.

Reclus' conception of love and solidarity is also relevant to issues in contemporary ecophilosophy. While various recent theorists have offered "identification" with nature as an antidote to anthropocentric attitudes and practices, such proposals have sometimes remained on a rather idealist level at which identification has the character of an act of will, if not that of a leap of faith. Reclus is closer to the position of social ecology and bioregionalism on this issue, as in many other areas. For him, it is our growing knowledge (in the sense of both *savoir*, understanding; and *connaître*, being acquainted with) of the earth and its human and nonhuman communities that offers an expanded scope for identification and solidarity. As we come to know each realm more adequately, we achieve greater identification with our own species, identification with all the inhabitants of the planet, and finally, as "the conscience of the earth," identification with the living, evolving planet itself. In this insight, Reclus anticipated some of the most profound dimensions of contemporary ecological thought.

4

A Philosophy of Progress

Although the myth of progress has taken on myriad forms over the ages, it has remained powerful through much of the history of Western civilization. Indeed, in various guises it has constituted the dominant myth of modernity. Even radical critics of existing society have had difficulty challenging it, and the classical anarchist thinkers, including Bakunin, Kropotkin, and Reclus, were no exception. Indeed, they sometimes rivaled their capitalist and statist opponents in their confidence in the inexorable advance toward a better future.[1] When one examines Reclus' view of history, one is struck by the strongly progressivist nature of his thought. In this, he seems to be quintessentially modern in his thought, imagination, and sensibility.

Nevertheless, Reclus distinguishes himself among classical radical theorists by the complex nature of his conception of progress. On the most overt level, he is a strong partisan of the concept and seeks to defend it against those who would use it on behalf of injustice and oppression. He recognizes that since the French Revolution the idea of progress has often been used as an ideological justification for elitism, class domination, imperialism, and other evils. Reclus attempts to rescue the concept from those who have betrayed it in this way, but on a deeper level, he questions the idea of progress itself. He refuses to interpret any given historical event, movement, institution, or tendency as being simply and unequivocally progressive. Instead, he takes a dialectical approach in which every historical phenomenon is seen as embodying in itself many contradictory moments. Thus, all such phenomena can be seen as having both progressive and regressive elements that require careful analysis if one is to understand their significance and assess accurately their dominant tendency.

One of the goals of Reclus' social geography is to uncover continuities between the natural and social worlds and between natural and social phenomena. His analysis of progress exemplifies this endeavor, for it seeks to show that the concept can be applied meaningfully not only to human society but also to the natural world. Just as in human history there are cases in which various "social types" have attained "full blossoming," so in nature there are examples of genera and species that "have reached such ideals of strength, rhythm, or beauty that nothing superior to them can be imagined." Thus "each form, epitomizing in itself all of the laws of the universe that converge to determine it, is an equally marvelous consequence of this process."[2] Reclus' analysis echoes such philosophical themes as Aristotle's concept of *telos* (according to which beings have ends toward which they aim and that, if attained, constitute the full realization of their highest potentialities); Aristotle's more specific idea of the *aretai*, or excellences, which define such elements of self-realization for human beings; and the Daoist concept of *de*, which refers to the power of beings to realize their unique and incomparable goods. Even more striking is the similarity between Reclus' thinking and concepts of recent ecophilosophers who focus on the importance of the fact that beings in nature have the capacity to attain their peculiar goods and that each has the possibility of developing into a "good of its kind." He sees such development toward a good as a key concept for understanding both nature and human society.

When Reclus looks at the vast scope of human history, he sees certain slowly developing but pervasive changes in society that are moving it toward a future in which it realizes its own good—that is, the attainment of freedom and justice in its institutions and practices. Although he argues for the need for periodical violent revolutions, he interprets such cataclysmic events as but the culmination of gradual changes that take place over long periods of time. Progress is thus the result of interdependent revolutionary and evolutionary processes. He offers as evidence of evolutionary change the slow decline in belief in certain scientific absurdities and religious superstitions, and the waning power of traditional hierarchical and deferential attitudes. He believes that contrary to some versions of historical materialism, changes in consciousness can precede and indeed give rise to fundamental changes in what is incorrectly seen as the "material base" of society. It is in part for this reason that he hopes "the great evolution now taking place" can prepare the way for "the long expected, great revolution."[3]

Though Reclus tries to show that world history (within which he would include so-called prehistory) contains both advances and regressions, he concludes that beneath these changes there is an overall movement of evolutionary progress. In interpreting social evolution, he sides ultimately with the view that "civilized" societies are more developed, not in the common economic or technical senses of this term but rather in relation to the project of human self-realization. He contends that while these societies have regressed in some important ways in comparison to previous ones, their advances make them on balance a positive step on the path of evolutionary social progress. Thus, the more "primitive" society often has the advantage of greater "coherence" and "consistency with its ideal," while the "civilized" one has gained in "complexity" and "diversity."[4] By this he means that the latter incorporates within itself a greater range of elements and has more highly developed interrelationships with other societies. For Reclus, the various forms of social "intermingling" characteristic of modern societies are enormous advantages since they expand our possibility of sharing human experience and attaining greater universality. In his view, the greater the interaction between cultures and races in a society, the greater the society's strength and ability to achieve the good for all its members. Consequently, the more that all cultures of the world are unified in a universal global society, the more the advances of every region and age can contribute to general human development.

Reclus' works sometimes exhibit a disconcerting juxtaposition of profound admiration for diverse cultures and condescension to those that he considers less advanced. Some of his earlier works in particular are lacking in appreciation of the merits of so-called "primitive" societies. Thus, writing in the 1860s in *The Earth*, he claims that in tropical regions "the mildness of the climate, the fertility of the soil, the exuberance of life, and the suddenness of death, take an equal part in maintaining man in his native carelessness and idleness," so that humans "bend in silence before the majesty of mighty nature" and reconcile themselves to being "her slave."[5] Like many of his contemporaries, he depicts such societies as remaining at the level of the "childhood" of humanity. He chides the Europeans for falsely taking pride in advances that were due less to their unique qualities and more to their good fortune in happening to live in the temperate zone, where they are "incited to labor" and pushed into efforts that led them to acquire "shrewdness, knowledge, cheerfulness, and love of life."[6]

In his later works, he comes to recognize that many of these same intellectual and personal qualities exist in one form or another in societies other than the naturally "favored" ones. Even so, he sometimes lapses into generalizations about the "backwardness" of peoples of the tropics. For example, he argues in *Man and the Earth* that while in such regions people can live with little effort, they do not "prosper" because "a purely vegetative existence is not conducive to the development of [man's] intelligence" and will not "render him master of the too-indulgent nature that surrounds him."[7] He reiterates his opinion that the conditions necessary for these achievements are present only in certain "regions of effort" that "are all situated in the northern temperate zone."[8]

Despite such questionable assessments of various cultures, Reclus manages to escape the widespread prejudice of his age according to which civilization constituted an unambiguous advance compared to previous social forms. Indeed, he explicitly attacks what he calls "chronocentric egoism," which holds "that contemporary civilization, as imperfect as it may be, is nevertheless the culminating state of humanity, and that by comparison, all past ages were barbaric."[9] In his view, societies go through periods of both progress and regression, and there is no single, unilinear path of world history by which to judge achievements. He criticizes European observers who "haughtily" dismiss all tribal cultures as merely "savage," when in fact these societies are at "distinct points" in social evolution and may either be in a vital and creative "state of becoming" or "on the road to decay and death."[10] This recognition of the specificity of social evolution within each society shows a relative openness to the unique qualities of those societies. Reclus concedes that societies that are often arrogantly excluded by European thought from world history are in fact "in history"—that is, they have their own history, which is comparable to the history of any other society.

Furthermore, despite Reclus' statements concerning the greater degree of progress in modern societies, his writings demonstrate considerable sensitivity to the values and achievements of premodern and non-Western societies. He admires much about tribal societies and considers the modern world to be quite inferior in many important areas. He cites, for example, the Aeta of the Philippines, whom he considers to be a model "in goodness, in spirit of justice, in rectitude of intention, and in reverence, in the truth of word and deed."[11] Similarly, he praises the Aleuts for many qualities, including their artistic achievements, their boat-building and sailing accomplishments, their achievement of "social equilibrium," and their peacefulness and amiability.[12]

Reclus avoids both the narrowness of provincialists, who are blind to the achievements of such societies, and the naïveté of primitivists, who often have limited knowledge of tribal peoples yet idealize them as "noble savages." The latter observers, by not applying critical judgment to a society's practices and by subordinating its lived reality to their illusions, express a subtle condescension toward the society. Reclus contends that it is necessary to recognize that some tribal societies have engaged in the most hideous rites of murder and ritual decapitation, and that even such generally admirable cultures as that of Tahiti have included some brutal and inhumane institutions.[13] Yet there is no need to idealize such societies to find merit in them, for when examined in the most realistic light they offer examples of some of the greatest achievements of human self-realization.

The fundamental challenge for society, according to Reclus, is to discover and develop fully every area in which humanity has progressed, while at the same time uncovering and negating every tendency toward regression. He contends that modern society has regressed in comparison to tribal society not only in the area of social solidarity, but also in its relationship to the natural world. His position on this issue is very similar to that of many contemporary social ecologists who concur with Reclus that human society has throughout history substituted one form of social hierarchy for another and has increasingly adopted an exploitative and destructive standpoint toward the natural world. The result is an intensifying contradiction between the possibilities created through social progress and the costs imposed on humanity and nature for its continuation. There is a growing need to resolve this contradiction through the destruction of the system of domination that divides human beings from one another and from nature. The attainment of this goal will permit a reappropriation of those valuable aspects of human society that have been sacrificed, including the communal sensibility and the respect for nature embodied in earlier social formations. While it is neither desirable nor even possible to replicate past social phenomena, knowledge of such past achievements offers inspiration for transforming our values and for expanding our vision of human possibilities in the future.

For Reclus, as for many modernist political radicals, humanity's success in the project of liberatory social transformation requires both the vanquishing of ignorance and prejudice and a further development of human rationality. He states that the true revolutionary must be not only "a man of sentiment" but also "a man of reason," uniting a strong feeling of human solidarity with precise knowledge of history, sociology, biology, and other fields, so that he can "incorporate his personal ideas into the generic whole of the human sciences."[14] Advances in critical historical rationality are particularly important to Reclus. Through the scientific knowledge of history, humanity can learn how to preserve all the gains of historical progress and reclaim what has been lost through all the regressions of the past. "Modern man must unite in his being all of the virtues of those who have preceded him on earth. Without giving up any of the great privileges that civilization has conferred on him, neither must he lose any of his ancient strength, nor allow himself to be surpassed by any savage in vigor, dexterity, or in knowledge of natural phenomena."[15] It is notable that despite his use of the now-pejorative term "savages," Reclus recognizes among the many virtues of tribal people a greater knowledge of nature, which is no small matter from his strongly naturalistic, ecological point of view.

Indeed, Reclus' overall judgment on the modern world is that despite enormous progress in many spheres, regression has taken place in some of the most essential areas. Modern society is superior in complexity since it has attained "a greater scope and constitutes a more heterogeneous organism through the successive assimilation of juxtaposed organisms."[16] But this growing unity is achieved through the creation of a nation-state "whose aim is preeminence over and even the absorption of other ethnic groups," so that the diversity of cultures existing in decentralized tribal societies is increasingly lost.[17] Here again, there is a strong parallel with more recent social ecological analysis of the dialectic underlying the history of civilization. While later social forms achieve greater complexity and even universality in their embodiment of a vast process of historical development, at the same time they move toward social simplification and the destruction of the wealth of cultural diversity, as a global monoculture is established by the global capitalist economic order and the global statist political order. There is a striking parallel and a real historical connection between this social simplification and the accompanying ecological simplification, alluded to by Reclus in his discussion of the Maori cited earlier.

Throughout Reclus' work one finds such a dialectical social critique, in which the coexisting progressive and regressive moments of historical phenomena are delineated. His examination of the history of religion presents one of the most comprehensive and detailed applications of such an analysis. Beginning with tribal religions and the rise of the major world religious traditions and extending the analysis to his own day, he traces the changing role of religion in various social systems and delineates its historically progressive and regressive dimensions.

For example, he analyzes the long evolution of the Hebrew God. Beginning as a "belligerent defender of the confines of his narrow homeland," Yahweh is gradually transformed into a rather distant and transcendent ruler of the earth, and finally into a god of compassion, as the sorrow of the people, "having renewed the nation, renewed its god also."[18] He explains that this transformation was a response to the experience of the Jewish people, specifically to its experience of war, plunder, and betrayal by its leaders. As a result of such ordeals, it began to conceive of the deity not as "the protector of the homeland, but rather as the representative of justice."[19] Religion thus reflected a process of moral education in society.

Reclus contends that in Israel a "moral revolution" took place, in which the Hebrew prophets expressed a vision of social justice the legacy of which continues in the demands of reformers and revolutionaries of modern times. Prophets such as Amos and Micah "expressed their disgust for religious formalities, spectacles, sacrifices and genuflections," and epitomized the essence of religion as "pure and simple morality, the practice of justice, and kindness."[20] They condemned war, looked forward to a future age of peace, and "dreamed of that universal fraternity of which we still dream today and which has fled from us like a mirage over the past two thousand years."[21] Having lost their own land, they "embraced in thought the whole of the universe" and looked forward to the day when all would be united "in the perfect consciousness of what is just and good."[22] Needless to say, this vision of justice and solidarity is of much more than historical interest to Reclus. For it is exactly this moral heritage of the Hebrew prophets, which was passed down to him by way of radical Protestantism, that lies at the core of his own anarchism.

Despite Reclus' admiration for the prophetic tradition, he subjects it to a critical, dialectical analysis and finds in it not only progressive but also strongly regressive aspects. While the prophets' message of justice constituted an enormous contribution to the history of liberation, it was also inextricably enmeshed in the system of historical domination. Consequently, its powerful message was available for use in legitimating and even strengthening that system. Because it was allied with an authoritarian monotheism that claimed "the certitude of knowing the only God, the absolute Master," the prophetic tradition also contributed to the creation of the theocratic state and the first "perfect religious intolerance" in history.[23]

Reclus is particularly astute in diagnosing the ways in which the metaphysical and moral insights of the founders of spiritual traditions have been turned into ideology at the service of power. His discussion of the history of Buddhism perhaps best illustrates this. He shows how the revolutionary implications of Shakyamuni Buddha's message were negated as his teachings were institutionalized in the form of an organized religion in alliance with political power. Reclus is among the few Western social and political theorists who have understood the challenge to all existing ideologies and institutions inherent in the Buddhist appeal to direct experience. He perhaps sees the affinities between his own critique of domination and belief in universal love, and the fundamental Buddhist teachings of nonattachment and compassion. In Reclus' words, the Buddha preached a "gospel of brotherhood" that proclaimed "no more kings, no more princes, no more bosses or judges, no more Brahmins or warriors, no more enemy castes that hate one another, but rather brothers, comrades, and companions who work together!" He goes so far as to say that in this original teaching "all hierarchy is abolished" and that "there is no role at all for authority."[24] In effect, the original Dharma was a form of anarchism.

Reclus explains that these revolutionary, anarchistic dimensions of Buddhism were destroyed as its social egalitarianism was given a purely moral or mystical interpretation, the Buddha was declared a god (or an avatar of Vishnu), and the Dharma was established as an official state religion. He notes the irony of the fact that the state reestablished the caste system, while official state proclamations continued to proclaim such Buddhist principles as "human fraternity and the necessity of spreading instruction to women and children as well as men."[25]

Reclus notes an even more extreme conflict between ideology and practice in the case of the Jains. The Jain religion is based on a feeling of unity with all of life and an ethics of nonviolence that extends even to other species (a view that today would be called "biocentric egalitarianism"). Yet the far-reaching social implications of such principles, which would certainly require the abolition of the state and other authoritarian institutions, were negated as Jain practice developed into an extreme and even fanatical obsession with avoidance of injury to various life forms. It became perhaps the only biocentrism in human history that took its principles to their logical conclusions (though, some might say, by reducing them to the absurd). The Jains adopted such extreme practices as filtering drinking water, breathing through a veil, and sweeping the ground before them as they walked in order to avoid destroying other life forms. However, their respect for life did not prevent them from "enriching themselves at the expense of the populace" so that they became "a fierce caste, composed of public enemies who were justly detested by the people."[26] Reclus points out that their respect for all life forms did not prevent them from becoming an elite group that exploited the masses.[27] He remarks that "such is the fate of religions: in becoming established, they negate their own starting-points, systematize their betrayal, and repudiate their own founders."[28]

Despite Reclus' harsh critique of institutionalized religion and his professed atheism and secularism, there is an implicit, but very significant, religious undercurrent in his own works. As has been mentioned, he sometimes writes in a rather pantheistic vein of the experience of nature as involving a loss of the ordinary sense of selfhood and a merging with the surrounding milieu. In some works he expresses not only an intense love of the natural world but something close to the experience of union with nature typical of nature mysticism.[29] Furthermore, at times he explicitly refers to his own philosophy as a kind of humanistic religion based on the pursuit of the good of the whole. When the journal *La Revue* proposed discussion of "morality without God," Reclus replied that "the public good, or in other words the happiness of all human beings, our brothers, will naturally become the special goal of our renewed existence" and that "we will thus have our religion, which, henceforth, will be in no way incompatible with reason, and this religion, which is moreover far from new and has always been practiced by the best people, includes everything good that was contained in the ancient religions."[30]

Reclus holds that this positive core of religion has been overwhelmingly betrayed by its institutionalized forms; nevertheless, he recognizes that it is still put into practice by the more enlightened and compassionate adherents of these traditions. He admits that there are tendencies within religion that are compatible with the social goals of anarchism, even when there are irreconcilable divergences on the level of beliefs. Thus, in his letter to M. Roth, a minister in Orthez, he says that although "there can be no agreement between Christians and anarchists, for all confusion of languages leads to a confusion of ideas," nevertheless, "as a Christian, you carry out your mission conscientiously. We anarchists know that all the heartfelt love that you have for your non-Christian friends hastens the coming of that great federation into which all men of good will, going beyond all churches, will enter, even if they be atheists like the Buddha."[31]

Despite his recognition of the progressive aspects of religion, Reclus sees its regressive aspects as by far the most significant ones in the modern world. He finds institutionalized religion to be primarily a force that perpetuates a past of ignorance and superstition, and that stands in the way of social progress. He juxtaposes it starkly with science, which he sees as a force for progress, enlightenment, and modernity. Reclus traces the origins of religion as a social force back to early societies in which the shaman was both a teacher who conveyed knowledge based on observation of the real world and also a priest who propagated fantasies concerning an illusory world. He contends that throughout history traditional worldviews have inherited the legacy of this original split and have consisted of an amalgam of myth and reality, truth and falsehood.[32]

Reclus observes that as knowledge of society and nature has progressed in the modern world, science and religion have increasingly diverged from one another. He sees the result of this divergence as "a distinct opposition, a relentless war, between science—that is, the objective search for truth—and the collection of feelings, beliefs and fetishistic vestiges that we call religion."[33] He believes that in this struggle science must ultimately triumph and reveal the dominant religious concepts to be relics of past ignorance and superstition.

Science thus plays a heroic role in history for Reclus. Though he finds elements of both progress and regression in the history of science, his account exhibits the almost boundless faith in science and technology that is so typical of classic modernity. He sometimes depicts scientific institutions not only as essential to all material progress but, even more, as the key to truth in all realms. The march of progress advances inexorably, banishing obsolete ideas and overcoming material barriers, and science is the instrument of its triumph. In his preface to the 1892 French edition of Kropotkin's *The Conquest of Bread*, Reclus goes so far as to say that he "professes a new faith" and that the object of that faith, "science," is the "ideal" upon which society must ultimately "model" itself.[34] How this "modeling" might be accomplished is not made entirely clear. However, what is apparent is Reclus' adherence to a variation of the myth of the Enlightenment, in which technique and reason march forward in tandem, as ignorance, superstition, and material scarcity are progressively vanquished.

As a result of this rather extreme historical optimism, Reclus sometimes exaggerates the possibilities for banishing ideology from the modern world. "As the worker believes no longer in miracles," he asks rhetorically, "can he be induced to believe in lies?"[35] Unfortunately, we have seen over the last century abundant evidence of the capacity of human beings—workers or otherwise—to delude themselves, whether to justify their hope in times of desperation or their cynicism in times of complacent satisfaction. Reclus vastly underestimates the need of human beings to create illusions to deal with the eternal problems of human existence: pain, suffering, death, losses of all kinds, the search for identity, the quest for meaning. Like almost all classical radical theorists, he has an inadequate grasp of some of the most important spiritual, existential, and psychological dimensions of the human condition. He also devotes little attention to the deep-seated human striving for power that has been explored by the Hegelian, Nietzschean, and Lacanian traditions but has been generally neglected by most radical social thought, including much of the anarchist tradition. Consequently, he fails to consider adequately the ways in which ideals like "anarchy," "communism," or even his cherished "brotherhood" and "solidarity," not to mention "science," "reason," and "progress," could themselves so easily be distorted ideologically.

But although Reclus sometimes lapses into naïve technological optimism and uncritical rationalism, his thought often transcends these tendencies. As will be discussed later, he includes an incisive critique of technology in his overall critique of domination. He sees the ultimate criterion for judging social progress to be neither technological development nor economic growth but rather the advancement of human social self-realization in harmony with the natural world. Furthermore, he rejects narrow views of this goal that would identify it with vastly increased productivity, material improvements, expansion of knowledge, or even the maximization of pleasure and happiness, as utilitarian ethics maintains, and as the conventional wisdom seemed increasingly inclined to hold even in his day.

In place of any sort of technological or economistic utopianism, Reclus develops a many-sided view of human self-realization that includes some of the goals mentioned but goes far beyond them. It consists, he says, of "a complete development of the individual," including such areas as "the improvement of the physical being in strength, beauty, grace, longevity, material enrichment, and increase of knowledge," in addition to "the perfecting of character, the becoming more noble, more generous, and more devoted."[36] Moreover, self-realization must be a social process in which "the progress of the individual merges with that of society, united by the force of an increasingly intimate solidarity."[37] Even in this brief summary of the nature of the good life, Reclus includes its physical, aesthetic, material, intellectual, moral, and social dimensions. Reclus' view has much in common with the Aristotelian eudemonistic ethic, both in his multifaceted conception of human self-realization and in his belief in the inseparable interconnection between the individual attainment of diverse virtues or excellences and the collective realization of a common good. Though Reclus' ethics can in many ways be looked upon as "an ethics of care," it might also be seen as a "left Aristotelianism," in that it broadens the concept of self-realization into a radically universal one. It is in some ways more radical than Marx's left Aristotelianism, not only in proposing a conception of human solidarity with more deeply communitarian dimensions but also in linking human self-realization to similar processes of unfolding and development on the part of other species and of the earth as a whole.

Reclus seems particularly Aristotelian when he defines the ultimate goal of progress as "happiness," in a very expansive sense. He explicitly rejects the individualistic or utilitarian conception of happiness as mere "personal enjoyment," and he redefines it to encompass a process of universal self-realization. Happiness, he says, "is true, deep, and complete only when it extends to the whole of humanity."[38] But this does not mean merely totaling up the sum of all individual satisfactions to produce the aggregate good for humanity as a whole. The attainment of happiness cannot be reduced (in the manner of individualistic ethical theories) to "a certain level of personal or collective existence," but rather includes a collective "consciousness of marching toward a well-defined goal" and a process of "directing the whole great human body toward the greatest good."[39] "Happiness" is thus a broad concept signifying participation in humanity's collective project of self-realization.

From Reclus' holistic ethical perspective, the concept of self-realization must also be extended beyond our own species. If we are truly to act as "nature becoming self-conscious," we must, in addition to pursuing the good of humanity, also contribute as much as is in our power to the good of all living beings. Accordingly, humanity has a wide-ranging moral responsibility that consists not only of negative duties to refrain from harming the natural world but also of positive duties to contribute actively to its flourishing. We have an obligation to "develop the continents, the seas, and the atmosphere that surrounds us; to 'cultivate our garden' on earth; to rearrange and regulate the environment in order to promote each individual plant, animal, and human life."[40] Such actions must be based on a growing awareness of our integral place in nature that leads us "to become fully conscious of our human solidarity, forming one body with the planet itself, and to take a sweeping view of our origins, our present, our immediate goal, and our distant ideal."[41]

Holistic thought is sometimes criticized for emphasizing the good of the whole to the detriment of that of the part. While examples of authoritarian and even "fascistic" holism can certainly be found, the term in no way implies domination or "totalization." Indeed, the most authentically holistic approaches recognize that the self-realization of the whole can only be attained through that of the parts, whether these may be the body and all its organs and faculties, in the case of holistic health, or the community and all its constituent groups and individuals, in the case of holistic social theory.[42] Reclus' social thought is instructive as an example of a critical holism that gives full recognition to the relative autonomy and integrity of individuals. It is thus, like any truly dialectical social ecology, a theory of unity-in-diversity. The nature of social progress cannot be understood merely through an analysis of the development of structures, institutions, or other social wholes but also requires careful attention to individuality and subjectivity.

According to Reclus, all "evolution in the existence of peoples" is the result of "individual effort."[43] It is true that he often explains various social phenomena as the result of the interaction between natural and social forces. Yet he reminds us that history cannot be reduced entirely to the dialectical interplay between objective conditions and that the human freedom to act creatively and to shape the future always exists, albeit within certain social and natural constraints. His overriding concern is to demonstrate that it is "the human person," rather than historical laws, institutions, or social forces, that is the "primary element of society."[44] Society, he believes, becomes more deterministic as economic power, the authoritarian state, and manipulative technologies subject people to more rigid controls and as freedom and creativity find fewer (or perhaps less conspicuous) avenues for social expression. Yet spontaneity and choice remain possible, and they constitute the basis for creating a future society in which nondominating unity-in-diversity is finally realized. "Free society establishes itself through the liberty provided for the full development of each human person, the original basic cell of society, who then joins together and associates as he wishes with other cells of a changing humanity."[45] Reclus' account of the achievement of personal self-realization through participation in the free, cooperative community is an excellent example of that "commitment to communal individuality" that Alan Ritter sees as "the strength of the anarchists' thought."[46]

Reclus' vision of the unfolding of freedom in human history continues in many ways the Spinozistic and Hegelian conceptions of freedom as a form of self-determination and self-expression. Spinoza defined the attainment of freedom by a being as its movement from being passively acted upon by external forces to shaping actively the world around it. Hegel developed and radicalized this conception by giving it a historical dimension. Spinoza's *Deus sive Natura* (God or Nature) becomes in Hegel *Geist* (Spirit) or the Absolute, the universal subject of world-historical development toward freedom through self-realization. Reclus' philosophy of progress is a major step toward naturalizing and demystifying such a conception of freedom. He notes that while "for a long time we were nothing more than [the earth's] unconscious products, we have become increasingly active agents in its history."[47] As humanity becomes more aware of its agency, it can develop a meaningful conception of its collective self-liberation. Thus, "the essence of human progress consists of the discovery of the totality of interests and wills common to all peoples; it is identical to solidarity."[48] Finally, our conception of unity will extend beyond our own species to the earth itself. As we begin to see ourselves as "one body with the planet" and conceive of ourselves as "nature becoming self-conscious," we realize that the process of attaining freedom through self-realization encompasses not only humanity but also the earth as a whole.

Needless to say, from Reclus' perspective, while this general progressive movement toward freedom takes place, contrary regressive developments are also impeding its advance. He believes that one cost of the development of civilization has been an increase in the barriers between individuals and groups in society resulting from institutionalized domination. Social progress therefore depends on the elimination of hierarchical divisions, so that open communication can be achieved. Long before Habermas, Reclus discussed the idea that social emancipation requires forms of communication that are free from domination, and the ability of the species to create a fund of practical knowledge relevant to that emancipation. Reclus asserts that it is necessary to destroy "the boundaries between castes, as well as between peoples,"[49] so that humanity will finally be able to draw on the experience of all cultures and all individuals in formulating its goals and values. As a result, humanity will be able to synthesize the virtues and achievements of both modern and "primitive" societies. The complexity, diversity, and universality of the former will be allied with the latter's "original simplicity" of life, in which there is "a complete and amicable freedom of human social intercourse."[50]

As humanity becomes in this way more conscious of its own history, it also expands its consciousness into the larger context of nature. The concept of humanity as the self-consciousness of nature was present in Reclus' work almost three decades before he made it the opening statement of *Man and the Earth*. As early as the 1860s he asserts that "since civilization has connected all the nations of the earth in one common humanity—since history has linked century to century—since astronomy and geology have enabled science to cast her retrospective glance on epochs thousands and thousands of years back, man has ceased to be an isolated being, and, if we may so speak, is no longer merely mortal: he has become the consciousness of the imperishable universe."[51]

Reclus had already begun to develop his dialectical conception that all phenomena of nature are in a constant state of transformation and unfolding. He notes that one of the most basic truths discovered through the universalization of consciousness is that "everything is changing and everything is in motion."[52] Humanity begins to understand itself in the context of these larger processes when it begins to look beyond human history to earth history. Thus, "the firm ground which he treads under his feet, and long thought to be immovable, is replete with vitality, and is actuated by incessant motion; the very mountains rise or sink; not only do the winds and ocean-currents circulate round the planet, but the continents themselves . . . are slowly traveling around the circle of the globe."[53] Over the subsequent decades, Reclus increasingly developed this process view of reality in a teleological (though in no way deterministic) direction as he united it with a vision of universal self-realization.

Reclus deserves recognition as an early prophet of the developing globalization or planetization of humanity. He believes that as the world is brought closer together through advances in transportation and communication, and as knowledge of the common history of human beings and the earth grows, humanity will have to revolutionize its system of values in accord with its growing unity. He notes that "industrial appliances, that by a single electric impulse make the same thought vibrate through far continents, have distanced by far our social morals";[54] further, the exploration of all corners of the earth and the perpetual movement of travelers between all countries has made us "citizens of the planet."[55] He concludes that through such means the illusion of separateness has become untenable and that "humanity has arrived at self-consciousness."[56] In the *New Universal Geography* he strikingly expresses this conception of the growing planetary unity of humanity:

> So bounded are now the confines of the planet, that it everywhere benefits by the same industrial appliances; that, thanks to a continuous network of postal and telegraphic services, it has been enriched by a nervous system for the interchange of thought; that it demands a

> common meridian and a common hour, while on all sides appear the inventors of a universal language. Despite the rancors fostered by war, despite hereditary hatreds, all mankind is becoming one. Whether our origin be one or manifold, this unity grows apace, daily assumes more of a quickening reality.[57]

Reclus thus saw the unification of humanity as a concrete, material process that will increasingly be experienced as a historical reality. In his view, a growing consciousness of this process will make the anarchist ideal of social unity-in-diversity appear increasingly plausible to humanity. History increasingly exhibits "a general tendency of things to merge themselves into one living body in which all the parts are in reciprocal interdependence," so that society can be seen as moving toward a state in everything "would constitute a harmonious cosmos in which each cell retains its individuality, corresponding to the free labor of each individual, and in which all would mesh together with one another, each one being necessary to the work of all."[58]

Of course, the consciousness of global unity that Reclus hoped for was in his own time, and largely remains today, in only a rudimentary state. Yet he would argue that social and technological developments have nevertheless created objective conditions that help form the basis for such a planetary consciousness. He does not underestimate the obstacles to overcoming ideological distortions of this consciousness and to transforming it into effective social praxis, yet he is able (perhaps in an act of modernist, progressivist faith) to see profound, indeed revolutionary implications in the slowly growing awareness of the interconnections between all terrestrial phenomena.

Reclus is often marvelously imaginative in attempting to contribute to the creation of this new, unified vision of the world. An example is his proposal that an enormous globe with a network of surrounding walkways be constructed at the center of Paris, so that people could pass at various levels examining the details of the earth and thereby begin to grasp it as a vast interconnected whole. He also proposes that a new calendar be adopted that would not be linked to the history of any particular religious tradition or show preference for one culture over another. He judges the idea of numbering years in two directions with positive and negative numbers to be completely irrational. His solution is to choose a beginning point with a universal, planetary significance rather than a merely particularistic, culturally specific one. He suggests for this point of reference the first eclipse recorded in human history. He notes that he would be writing in the year 13,447 according to this system.[59] His choice of the first recorded eclipse is a powerful symbol of the interrelation between the natural and the social. Although that event was a natural occurrence beyond human control and involved phenomena extending even beyond our own planet, as a recorded event it forms part of human history and is noteworthy for its place in the development of human knowledge of natural phenomena.

Reclus' desire to promote the unity of humanity seems sometimes to go to extreme lengths, as when he discusses the need for a "common language" for all human beings to facilitate global communication. He certainly appears a bit naïve when he states that the members of the new nation of humanity "must understand each other completely" and suggests that a language might be developed that would help realize this rather ambitious goal.[60] However, there may be more value to his speculation on this topic than appears at first sight. Reclus was well aware of the historical importance of Latin and later French as common languages of politics, commerce, culture, and scholarship. Today he would no doubt point to the growing dominance of English in these areas and in commerce as the expression of the need for an ever more closely interrelated humanity to express itself in a common tongue. Clearly, he would have preferred Esperanto or a new, more multicultural and universalistic language for such a means of communication. Yet the fact that far more human beings than ever before can now communicate directly would be seen by Reclus as strong evidence of progress in the unification of humanity. He would certainly add that the inevitable regressive dimensions of such a development should not be ignored, no doubt pointing out the cultural homogenization and loss of diversity that has accompanied the growing dominance of English and Anglophone culture.

For Reclus, the self-consciousness of humanity will continue to grow as knowledge of geography and history create a new global spatiality and temporality. In his view, "humankind, which makes itself One at every latitude and longitude, similarly tries to realize itself through one form that encompasses all ages."[61] He posits a close relationship between this growing knowledge (which Hegel calls "world-historical" but which in the spirit of Reclus we might call "earth-historical") and the expansion of freedom. For Reclus, history exhibits a certain order and "logic of events," the knowledge of which allows us to play a more active and creative role in determining our own destiny. To the degree that we broaden our grasp of historical development and the diversity of human possibilities, we transcend any "strict line of development" determined by environmental and hereditary factors.[62]

The achievement of a grasp of this historical unity-in-diversity is the goal of Reclus' social geography, especially as expressed in his sweeping account of human history in *Man and the Earth*. There he surveys the vast diversity of human achievements, while at the same time synthesizing that multiplicity into a single narrative of the human struggle for self-realization. He believes that such a unifying narrative is increasingly inscribing itself in history as a social reality rooted in human experience. As we become increasingly planetary beings in many spheres of our activity, social phenomena naturally begin to appear to us as aspects of the life story of a universal humanity. Cultures of other times and other places lose their quality of alien otherness, and their contributions to progress become available to all as examples of human possibilities. We come to see all peoples as "brothers toward whom we feel a growing spirit of solidarity," and we find throughout history "an increasing number of models demanding understanding, including many that awaken in us the ambition to imitate some aspect of their ideal."[63] The human race discovers itself to have a common history and is able to undertake a common project of self-realization. For Reclus, this means that the diverse experiences of all become part of one great human experiment, the great struggle for the attainment of freedom.

5

Anarchism and Social Transformation

Reclus is the anarchist geographer par excellence. The term "anarchist geography" captures perfectly the idea of his work: writing (*graphein*) the history of the struggle to free the earth (*Gaia*) from domination (*archein*). Yves Lacoste calls the work of Reclus, and above all his book *Man and the Earth*, the "epistemological moment," indeed the "epistemological turning point," in the history of geography. Before Reclus, he says, geography "was linked essentially to the state apparatus, not only as a tool of power, but also as an ideological and propagandistic representation. Reclus turned this tool against the state apparatus, the oppressors and the dominant classes."[1] For Reclus, social geography and the social philosophy grounded in it become part of the process of the planetary history of liberation.

It is reported that Reclus once exclaimed to the Dutch anarchist Ferdinand Domela Nieuwenhuis, "Yes, I am a geographer, but above all I am an anarchist."[2] Quite early in life he developed a deep faith in human freedom and solidarity that increasingly defined his existence and later received full development in his libertarian political theory. His anarchist vision of social freedom is also the mature expression of his enduring belief in moral autonomy. For Reclus, and as anarchist ethics from William Godwin on has so often stressed, moral responsibility is impossible without moral autonomy. In his early manuscript "Development of Liberty in the World," he asserts that "laws must appear before the tribunal of our conscience and we must not submit to them except when they are in perfect accord with the moral law that dwells within us."[3] If these laws conflict with "eternal justice," it is our moral obligation to disobey them. Respect for human laws in disregard for the higher moral law is no virtue and indeed amounts to no more than "moral cowardice."[4] While Reclus later dropped the rather abstract, idealist language of "moral law" and "eternal justice" in favor of a more historical and naturalistic depiction of morality, an emphasis on free commitment to the greater good of humanity and nature remained fundamental to his anarchism.

For Reclus, though "anarchy," aims at the greatest possible realization of freedom and justice and the establishment of a universal community based on freedom, justice, solidarity and love, it is never merely a vague and distant future utopia. It is capable of immediate realization wherever these values are embodied in existing human relationships and social practice. "Anarchy" is the entire sphere of human life that takes place outside the boundaries of *arche*, or domination. He states in the preface to the 1892 French edition of Kropotkin's *The Conquest of Bread* that "anarchistic society has long been in a process of rapid development," for it can be found "wherever free thought breaks loose from the chains of dogma; wherever the spirit of inquiry rejects the old formulas; wherever the human will asserts itself through independent actions; wherever honest people, rebelling against all enforced discipline, join freely together in order to educate themselves, and to reclaim, without any master, their share of life, and the complete satisfaction of their needs."[5] In effect, the entire history of the struggle for human collective self-realization constitutes anarchy, though it is often "unaware of itself."[6]

Although "anarchy" thus has a larger historical meaning for Reclus, he also uses the term to refer more specifically to a future society that is free from institutionalized forms of domination and that will attain an unprecedented synthesis of liberty, equality, and community. For Reclus, as for the anarchist tradition in general, anarchism means much more than anti-statism, opposition to coercion, or rebellion against authority. In its most sophisticated forms, it proposes a practice of social transformation and reorganization based on nondominating mutual aid and cooperation.[7] Reclus believes that the most deeply rooted social order arises out of the greatest possible freedom and voluntary association, and that ever-growing social disorder results from coercion, oppression, and domination. Thus, he sees "an abyss between two kinds of society," one of which is "constituted freely by men of good will, based on a consideration of their common interests," whereas the other "accepts the existence of either temporary or permanent masters to whom [its members] owe obedience."[8] The former has "authentic organization, spontaneous, attractive association that constantly adapts to the changes in persons and things," while the latter consists of "a forced juxtaposition that is opposed by continual tendencies to disjoin the parts. The former . . . is precisely the kind that by the very fact of its liberty remains centripetal, while the latter, held together only by regulations, is made up of centrifugal elements."[9] Reclus' goal is to develop a positive vision of such a future society of "ordered anarchy."[10]

However, in examining Reclus' reflections on the nature of anarchist society, one is often struck by the generality of many of his statements and the lack of specific content. In this, he is typical of the classical anarchist theorists and rather different from utopian writers, who often present highly imaginative depictions of a free and just society. Anarchy constitutes for him an inspiring social ideal that could give direction to present-day struggles; however, the details of future social organization must be arranged "after the revolution." In his essay "Anarchy," he summarizes the ideal as "equality of rights and reciprocity of services," and the basis of anarchist morality as the familiar principle "to each according to his needs, from each according to his powers."[11] The achievement of anarchy thus means simply the creation of a free, egalitarian, and cooperative society to replace the existing oppressive, hierarchical, and competitive one.

Reclus does at times discuss in general terms some of the institutions that might exist after the social revolution. At the Berne Congress of the League for Peace and Freedom in 1868, he proposes that a future society should be one in which all previously existing political divisions are replaced by workers' associations. In his view, the existing subdivisions, from the province down to the local district, are nothing but "tools of despotism" created by those who wish to centralize power.[12] He goes so far as to say that there is no such thing as a "natural border,"[13] for natural features must be given a social meaning through human action. Free individuals, he argues, will look beyond all artificial territorial boundaries and achieve "ideal justice" by reorganizing society through "productive associations and groups formed by these associations."[14] The boundaries of the free associations of the future may or may not correspond to existing borders, depending on the decisions of their members in their pursuit of justice and mutual aid. "Anarchy" in the sense of a fully realized anarchist society will consist of a large voluntary federation of these free associations existing at the local level.

In Reclus' view, such a free, cooperative society can only emerge out of a social revolution, but this revolution will itself depend on a long history of liberatory thought and practice. He observes that anarchism has spread "where minds have long been liberated from religious and monarchical prejudices, where revolutionary precedents have shaken faith in the established order, where the practice of municipal liberties has best prepared men to become their own masters, where disinterested study has developed thinkers free from all sectarianism."[15] There are thus a great many spheres of thought and action in which anarchists can contribute to social progress and lay the groundwork for the future libertarian society. In order for successful *revolutionary* change to take place, a long history of *evolutionary* change must prepare the way.

Reclus distinguishes himself from his historically more influential ally Bakunin in his deeper analysis of the preconditions for social transformation. While Bakunin made an important contribution to a critical theory of libertarian social transformation, he also succumbed to a fetishism of revolution and often exaggerated the liberatory potential of reactive social movements, vague popular discontent, and unfocused rebellion. For Bakunin, such amorphous social conditions could be given a revolutionary direction when shaped by an "invisible dictatorship" of conscious revolutionaries. Such views led to an exaggerated emphasis on revolutionary will and a vanguardism that has decidedly nonanarchistic and, indeed, authoritarian implications. The historical anarchist movement has often been influenced by a Bakuninist insurrectionism, and it might have done well to follow consistently Reclus' more balanced view of the relationship between evolution and revolution.

Reclus' idea of the complementary roles of evolution and revolution is one of the central themes in his political writings. "In every sphere," he says, "we are not only evolutionists, but also just as much revolutionists, since we realize that history itself is but a series of achievements that follows a series of preparations. The great intellectual evolution that emancipates minds has a logical consequence in the emancipation of individuals in all of their relationships with other individuals."[16] The potential contribution of any phenomenon to evolutionary change is not easy to ascertain, according to Reclus, because all phenomena have both positive and negative moments, and both progressive and regressive aspects in relation to the larger milieu. Each phenomenon is "two-sided, for it is at once a phenomenon of death and a phenomenon of revival; in other words, it is the result of evolution toward decay and also toward progress."[17] The challenge to those with a critical faith in historical progress is to preserve and develop the positive moment while rejecting and eliminating the negative.

Reclus applies this analysis to specific social institutions and phenomena. For example, he has enormous confidence that many advances of modern science and technology can be used for such progressive purposes as the increase of knowledge, freedom, health, and beauty, but nevertheless he also sees within them the potential for unprecedented levels of regimentation, domination, malaise, and degradation of society and nature. This forthright recognition of the dual nature of social realities distinguishes Reclus from many other modernist thinkers of his age, who focused one-sidedly on the possibilities for progress but neglected the dangers, costs, and self-contradictions of seemingly progressive historical developments.

From Reclus' dialectical perspective, revolution itself partakes of the dual nature characteristic of all social phenomena. There is no absolute revolutionary break with history, as idealist and voluntaristic revolutionary theorists would contend. Revolution is an integral part of the movement of history and reflects the complexity and contradictory nature of all the other historical phenomena that interact with it and condition it. Reclus points out that although a given revolutionary movement may be authentically liberatory in many ways, the revolutionaries have been shaped by the conditions existing prior to the revolution. These conditions do not disappear absolutely on the great day of revolt but rather leave traces on the personalities, practices, and institutions of the relatively transformed society.

Consequently, the exercise of revolutionary power often becomes a convenient tool of aspiring authoritarians, who transform revolutionary ideals into authoritarian ideology. In Reclus' words, "there is often a most shocking disparity between the revolutionary circumstances that accompany the emergence of an institution and the manner in which it functions, which is completely opposed to the ideals of its naïve founders."[18] In a prescient commentary on many later revolutionary regimes, Reclus notes the danger of "the routine, the hierarchy, and the spirit of regression that gradually encroach on every institution"[19] once a new system of concentrated power is established.

These reservations did not, however, deter Reclus from actively supporting revolutionary movements and seeking to help them transcend their limitations. He was dedicated to the First International, which he saw as an advance of historic dimensions in the direction of unifying humanity for the cause of justice and progress. He contends that "since the discovery of America and the circumnavigation of the earth, no achievement was more important in the history of man," and that "the future normative unity that the philosophers desired" only began to be realized when "the English, French, and German workers, forgetting their different origins and understanding one another in spite of their diversity of languages, joined together to form a single nation, in defiance of all their respective governments."[20] Although the practical results of that organization might hardly seem to justify such extravagant claims for its importance, his point relates more to its symbolic significance as the first modern embodiment in practice of the ideal of the unity of all of humanity and the example that it created for future efforts at global solidarity. In a sense, Reclus was saying that the First International ushered in the still rather incipient movement of "globalization from below."

Reclus' views concerning social transformation were profoundly affected by his participation in the International and by the influence of his close ally Bakunin. While Reclus and Bakunin opposed one another at various times on some important issues, including the role of secret societies in the revolutionary movement, the influence of the charismatic revolutionary was responsible in part for Reclus' development of a firm belief in the necessity of social revolution. He participated in the Bakuninist Alliance for Social Democracy and in Bakunin's efforts to move the nonrevolutionary League for Peace and Freedom in a more radical direction. He was also a member of Bakunin's International Brotherhood (a secret society of dedicated Bakuninist revolutionaries) from 1865 on. He attended the meetings of the General Council of the First International in 1869 and defended the anarchist (majority) position in the world's first great working-class organization.

A strategy of the international workers' movement that Reclus enthusiastically supported, and for which he had great hopes, was the general strike. He contends that "English, Belgian, French, German, American, and Australian wage workers understand that it is up to them to withhold all labor from their bosses on the same day," and he asks why they would "not carry out tomorrow what they understand today, especially if a soldiers' strike is added to that of the workers?"[21] From this passage it would seem that Reclus believed that the working classes of the major industrial countries of his time both understood clearly the strategy of the general strike and were committed to it in principle. If these two assumptions were indeed correct, it was reasonable for him to hope that a revolutionary situation on an international scale was imminent.

Unfortunately, such expectations exhibit some of the same kind of unrealistic revolutionary optimism that plagued Bakunin. While Reclus was right about the general strike not being in principle impossible, he overestimated the existing level of consciousness of the European working class. The kind of the careful analysis he applied to other issues might usefully have been devoted to the nature of the barriers confronting the expansion of popular critical consciousness. Although he has some important insights in this area, especially interspersed among his geographical writings, in his most explicitly political (and most widely reprinted) works, edifying revolutionary rhetoric often takes the place of probing analysis of the actual state of workers' consciousness, of the material, ideological, and imaginary processes that shape that consciousness, and of the factors that might transform it into an effective revolutionary force.[22]

Reclus' assessment of "propaganda by the deed," a subject of much controversy within the anarchist movement of his day, also presents certain problems. During the 1880s and 1890s, attacks on political officials, bankers, and industrialists, and even random victims in places judged "bourgeois," became increasingly common. The names of terrorists like Ravachol, Vaillant, and Henry became well known to the public. Many of those carrying out violent attacks on the established order began to invoke anarchist principles to defend their deeds, causing a crisis of conscience for anarchist theorists. Such actions would seem to be in direct contradiction to Reclus' ethical values and his humanitarian sensibilities. Indeed, in a letter of 1883, he asserts that "from the revolutionary point of view, I am very careful not to recommend violence, and I am distressed when friends carried away by passion allow themselves to resort to the idea of vengeance, which is so unscientific and sterile."[23] In another letter of 1889 he states that "the secret" is "to love everyone always, including even those whom one must fight with unflagging energy because they live as parasites on the social body."[24] And finally, in a letter of 1903 he goes so far as to assert that "it is necessary to resist evil without hating the evildoers, but rather even while loving them." Reclus was consistent in his rejection of hatred and vindictiveness both as a matter of general principle and in his own personal practice.[25]

Nevertheless, his refusal to advocate violence and vengeance did not in his mind require condemnation of those who come to see individual violence as a legitimate response to oppression and who act on these views. Although some well-known anarchists disassociated themselves from all terrorist acts and others, like Kropotkin, adopted an ambiguous position, Reclus steadfastly refused to condemn the propagandists of the deed. In his opinion, violence in society is the necessary result of a cruel and inhumane system of oppression, and blame should not be directed at those victims who in desperation lash out against their own oppression. Rather, in his view, those who control the unjust system and benefit from it should be held guilty for both the injustices that they inflict on society as a whole and the violent acts to which they drive some of the oppressed.

At times, Reclus came even closer to explicit approval of terrorist acts. In a letter of 1892, after expressing admiration for Ravachol for his "high character," he states that he considers "all revolt against oppression to be a good and just act."[26] Though he reaffirms his belief in the gradual growth of enlightenment through the effect of "words and feelings," he notes that anger "has its raison d'être, its day and its hour."[27] In another letter of the same month, he praises Ravachol for "his courage, his goodness, his greatness of soul, and the generosity with which he pardons his enemies, and indeed those who informed on him." He says that he knows "of few men who surpass him in nobility" and judges him "a hero of uncommon magnanimity."[28]

Reclus' considered opinion of all social phenomena in general is that they should be assessed carefully for their positive and negative aspects and that their effect on the overall course of social progress should be the final criterion for judgment. However, in the case of propaganda by the deed, he veers in a strongly deontological direction. In this instance, he stresses that the perpetrators of such acts should be judged by the nobility of their actions—perhaps even by the beauty of their souls—rather than strictly by their effects on the course of history and on the revolutionary movement they claimed to represent.

Reclus overlooks a number of crucial points concerning propaganda of the deed. First, the deterministic arguments that he invokes in order to excuse the terrorists have certain implications that he ignores. By the same reasoning, he should have focused more on the innocence of the terrorists' victims, for to whatever degree they participated in an unjust social system, they did not personally and intentionally create it and were certainly themselves products of that system. Second, whatever determinants may have been present, his refusal to hold the terrorists responsible for their actions denies them the status of moral agents capable of choosing between alternative methods of protesting against injustice. Instead, they are treated as no more than links in a chain of causality. Finally, the acts of desperation that the terrorists committed were, in any case, miserable failures that did little promote authentic social transformation and often only contributed to promoting reaction and repression. As "propaganda" they were a disaster for which anarchists are still unjustly suffering, stereotyped as they are as terrorists and "bomb-throwers."[29]

However, despite the many problems with his position on this issue, Reclus presents one quite powerful argument that should not be overlooked. He points out that to condemn the relatively rare violent acts of desperate individuals crying out for justice while at the same time complacently accepting the enormous system of day-to-day violence embodied in such social institutions as state domination, capitalist exploitation, institutionalized racism, and patriarchal oppression constitutes the worst form of ideological distortion. It is his concern for the widespread moral insensitivity to the horrors of entrenched, institutionalized injustice that leads Reclus to emphasize terrorism as a symptom of greater evils, rather than as an evil in itself.

Another of Reclus' most controversial views is his acceptance of the right of the workers "to partial recovery of the collective products" of society by means of the individual's "personal recovery of his part" of that property.[30] He means by this the sort of activity that is usually labeled "theft." Reclus believes that theft, like violence, is a great evil. But in his opinion, those who are outraged by this evil should direct their indignation above all toward the capitalist, statist system that institutionalizes theft rather than toward exploited individuals who informally use theft as a means of striking back at that system. According to his analysis, the thief is a "restorer" who seeks to reappropriate a small part of the wealth that has been extracted from his or her labor. In a letter of 1887 he writes that since private property is itself theft, "if a repossessor infringes on it, inspired by a spirit of justice and solidarity, I find no fault with it."[31] He adds that he is not inclined "by nature, by habit, or by personal tendency" to act similarly but that he has no right to "speak as a model for others."[32]

While some were shocked by this gentle man's advocacy of such activity, he argues that their horror is misplaced. Why, he asks, should we echo the dominant culture's hypocritical condemnation of the efforts of the oppressed to improve their miserable position in society through such reappropriation? First, he argues that under the existing exploitative order, theft is universal. He explains in a letter to Jean Grave that "in the society of injustice and caprice in which we live, we are, in spite of ourselves, implicated in all the evil that takes place."[33] And in a letter to the anarchist journal *La Révolte*, he asserts that "it is true beyond doubt that in this iniquitous society in which everything rests on inequality and hoarding, in which money alone provides one's bread, we are all, without exception, forced by the very conditions of our existence into a life of outright theft."[34] The truly abhorrent form of theft is that practiced by the rich and powerful, who are highly successful in their efforts to confiscate the product of the labor of others. Reclus' sense of justice was outraged by a system that exalts the biggest and most successful thieves and holds them up as models of virtue and of success in life while condemning to misery, disdain, and imprisonment those who seek to recover at best a small portion of what has been stolen from them.

There are some obvious problems with this analysis. First, there is the troubling question of the possible corrupting effects of this "reappropriation" process on those who carry it out in sovereign moral isolation.[35] While his communitarian ideas imply a collective process of seeking justice, he apparently sees the "restorers" as justified in appropriating individually what they personally believe to be due to them. To expect objectivity in such a process seems unrealistic at best. Furthermore, Reclus' contention that so-called theft should not be condemned because everything is theft has rather disturbing implications. On the one hand, it demonstrates an awareness of the manner in which all become implicated in systems of domination and injustice. On the other hand, it implies a moral equivalency for all actions "before the revolution" that threatens to create a nihilistic, rather than an anarchistic, ethos. If everything is theft, everything is deceit, and everything is exploitation (since we participate in corrupt systems in which these evils are ubiquitous), then "everything is permitted." However high Reclus' own moral standards may have been, he advocates on this issue a kind of moral laissez-faire that might justify egoistic self-interest as effectively as it would inspire liberatory social practice.

It should be stressed that Reclus' concept of anarchist politics does not focus on such isolated acts of defiance. Rather, it overwhelmingly emphasizes the importance of collective and communal organization and the growth of a culture of freedom and solidarity. He undertakes an investigation of the history of libertarian and communitarian achievements going back as far as the Athenian polis and ancient tribal societies, presenting an imaginative vision of the possibilities for embodying this experience in a transformed society. He wishes to reclaim the history of free community over the ages and to show how this tradition can be reinvigorated through the creation of a new libertarian and communitarian society.

Reclus differs markedly from other radical political theorists of his time in his claim that many elements of this long history were of more practical significance than the prevailing strategies of his own era. In his view, "the names of the Spanish comuneros, of the French communes, of the English yeomen, of the free cities in Germany, of the Republic of Novgorod and of the marvelous communities of Italy must be, with us Anarchists, household words: never was civilized humanity nearer to real Anarchy than it was in certain phases of the communal history of Florence and Nuremberg."[36] Despite his strong commitment to the contemporary workers' movement, he refused to narrow his vision of social transformation by limiting it to the model of the struggle of labor against the modern state and capitalism. For him, humanity must self-consciously seek self-realization by drawing on its long and expansive history of struggles for liberation and experiments in freedom.

Reclus attributes special significance in the history of human emancipation to the Athenian polis and to the achievements of Greek democracy.[37] He notes that in the polis, "the political unity [*ensemble politique*] of the social body was as simple, as undivided and as well-defined as was the unity of the individual himself," and that it "is in this sense that one must, like Aristotle, consider the human being to be par excellence the *zoon politikon:* the 'urban animal,' the participant [*le part-prenant*] in the organic city [*la cité organique*] (and not merely the 'political animal,' as it is usually translated)."[38] Thus, the "political animality" of the citizens does not mean merely that they were socialized or educated to possess "civic virtue" or that they achieved self-realization through the political community (though it certainly encompasses both of these). Reclus stresses the more holistic dimension of Aristotle's conception. When a being attains its end (*telos*) within a larger whole, it is an organic part of that larger whole. However, the citizens are not mere structural cells or organs in the body politic but rather dynamic *participants* in the larger organic unity. Reclus' use of the language both of life [*organique*] and of social action [*prendre part*] is significant. The free and democratic political community is a unity in which organic solidarity and spontaneous, voluntary activity are synthesized.

Reclus situates the development of the democratic polis within the larger scope of Greek history. He notes that the age of democracy coincided with a growing economic equality, as the aristocratic landowners lost some of their holdings and the lower classes of citizens gained wealth through the vicissitudes of war. Many of the old prerogatives were eliminated, and positions were increasingly opened to the electors. Political democracy thus coincided with other important economic and social reforms. The political, in turn, accelerated changes in other spheres. Reclus attributes the success of the Athenians in trade and commerce and the vast achievements of Greek culture in this period to the effects of growing political equality and democratization, which fostered creativity and initiative in all areas.[39]

Another epoch that Reclus recognizes as a milestone in the history of human liberation is early Icelandic democracy. He greatly admires the Icelanders for exhibiting a spirit of independence and for creating strong democratic traditions during a period in which Europe was mired in monarchical despotism and feudal hierarchy. He claims that they "succeeded completely in maintaining their dignity as free men, without kings, feudal princes, hierarchy or any military establishment."[40] Instead, they made decisions through a process in which "the common interest was discussed in the open air by all the inhabitants, who were dressed in armor, the symbol of the absolute right of personal self-defense belonging to each individual."[41] These assemblies took place at a volcanic gorge called the Almannagja, or "the Gorge of All the People," where the Lögmadr, or "Reader of the Law," proclaimed the decisions of previous assemblies. Any decisions that were not announced and reaffirmed by the assembly for three successive years were annulled. Reclus notes that judicial processes were subject to the same popular supervision. At the end of the gorge was "the Mound of the Law," where "the judge and the accused met face to face, under the vigilant eye of the armed multitude."[42]

Like Kropotkin, Reclus also looks to the era of the medieval free cities and their federations for inspiration for future social transformation.[43] He notes that these cities had two principles of association—one grouping citizens according to "professional interests, ideas, and pleasure," the other according to "neighborhood, district, and small territorial units that were supposed to be in no way sacrificed for the sake of the city center."[44] Rural communities had a similar dual organization, and both "joined together in leagues," some of which endured for hundreds of years.[45] For Reclus this presents an admirable model for future federations of local communities.

Despite the great achievements of medieval cities, Reclus also finds them instructive because of their shortcomings. He argues that a weakness of these communes was their lack of sufficient concern for the liberty of other communities, and their tendency to become absorbed with their own interests. As a result, they were susceptible to destruction by powerful economic and political interests.[46] He concludes that anarchists must devote as much attention to such principles as federation, mutual aid, and solidarity as they do to the goals of freedom, justice, and decentralization of power, if efforts to transform the larger society are to succeed.

Reclus also expresses great admiration for the independent Basque communities that "retained for centuries their administrative autonomy."[47] He comments that these communities have been distinguished by their love of freedom and their hostility toward all centralized authority. Since Reclus grew up in the southwest of France (Orthez is in the shadow of the Pyrenees), he knew the Basque culture well and found in it many affinities with his own anarchist sensibilities. He remarks that the Basques "have always preferred to live in isolation at some beautiful site in their land of hills and mountains, in the shadow of a great oak, symbolizing the tribe and its ancient liberty."[48] He takes this phenomenon as a good example of the interaction between aspects of culture and nature. In his view, the natural milieu encouraged the group's quest for cultural autonomy and individual freedom. He asks, "Where does the Basques' fine confidence in themselves come from, if not from that nature which has always protected them?"[49] Reclus is not implying any geographical determinism here, for he also notes the fact that the Basques knowingly sought out those aspects of nature that resonated most with their own yearning for independence. He is pointing out the sort of dialectical interaction between natural and cultural factors that can play a part in the quest for human liberation.

Reclus expresses a similar admiration for communities that had preserved traditions of communal property and cooperative labor. He celebrates the fact that there was in his day still abundant evidence of "the spirit of full association" in Switzerland, where "two-thirds of the alpine prairies and forests belong to the communes, which also own peat bogs, reed marshes, and quarries, as well as fields, orchards, and vineyards."[50] He describes these communities as exhibiting a joy in collective labor that contrasts markedly with the degraded conditions of contemporary industrialized manufacture and agriculture. He observes that "on many occasions when the co-proprietors of the commune have to work together, they feel as though they are at a festival rather than at work."[51] These small communities are important to him as evidence that the members of society can still work together cooperatively and that social organization based on solidarity is a viable alternative to the political and economic regimentation of an atomized society. As Reclus states it, such social cooperation persists "despite all the ill will of the rich and the state, who have every interest in breaking apart these tightly bound bundles of resistance to their greed or power and who attempt to reduce society to a collection of isolated individuals."[52]

It is not surprising that as Reclus surveys European history, he finds the French Revolution to be a great chapter in the history of human liberation.[53] It is noteworthy, however, that he also discovers that even this great landmark in the progress of humanity exhibits in a striking manner the dual aspect that he finds in all historical phenomena. On the one hand, it was a progressive step away from the absolute monarchy, the religious authoritarianism, and the cultural conservatism of the past. But it was also a regressive step in the direction of the authoritarian state and the concentration of political power.

For Reclus, the political unification and centralization that emerged from the Revolution were achieved at the expense of local liberties, and many of the most progressive features of traditional French society were destroyed: "Thus, the free communities, the 'universities' of the mountain people, lost their uncontrolled management of their own interests and their sovereign assemblies, in which each man and woman had the absolute right to presence, speech, and initiative."[54] The traditional leftist interpretation of the revolution has been that it was unequivocally progressive in destroying the traditional feudal and monarchical society and creating a republican system founded on the rights of the citizen—albeit in the form of a bourgeois republic. To Reclus, it is equally important to see how the revolution helped establish the modern nation-state that has progressively annihilated an invaluable legacy of decentralized, communal institutions.

An important epoch in the history of liberation that Reclus witnessed firsthand, and to which he personally contributed, was the Paris Commune. Not only did he spend years of his life in prison and then in exile as a result of defending it, but he also pondered deeply its political significance. He looked to this great social experiment as evidence of the growing historical efficacy of the principles proclaimed by the First International and the anarchist movement, but he also criticized the Commune for failing to live up to some of its own ideals. His critique is fundamentally a radically democratic, municipalist, and federalist one. In his view, "the principal error of the Commune, an unavoidable error, since it derived from the very principle on which power was constituted, was precisely that of being a government, and of substituting itself for the people by force of circumstances."[55] Its flaw was that rather than becoming a fully developed experiment in municipal liberty, and thus a model for other such experiments, it began to reinstitute the form of the state. The problem was that the necessary "evolution" had not taken place prior to the "revolution." The revolutionaries were still too much under the influence of traditional centralist, authoritarian politics to create a new, radically libertarian regime. "The natural functioning and intoxication of power led it to consider itself a bit like the representative of the entire French state, of the entire Republic, and not simply the Commune of Paris calling for a free association with other communes, towns and rural areas."[56]

Another expression of Reclus' radical municipalist outlook is his admiration for the great cities of history, his appreciation of the ethos of each city as a unique cultural expression. "Each city," he says, "has its unique individuality, its own life, its own countenance, tragic and sorrowful in some cases, joyful and lively in others."[57] Such a generalized depiction of the city only begins to capture the full complexity of the urban milieu, for the city constitutes not only a distinctive whole but also a sum of distinctive parts. It must be understood both as "a collective personality" and also as "a very complex individual" in which each neighborhood "is distinguished from the others by its own particular nature."[58] The culture of cities thus exemplifies the concept of a dialectical interrelationship between unity and diversity.

Reclus believes that although the city can be studied as a distinct social phenomenon, the life of cities and urban problems can be understood only in relation to the institutional structure of society as a whole. As he puts it, the "urban question" cannot be separated from the larger "social question." As long as a system of economic injustice and political domination exists, cities cannot develop freely and fulfill their destiny as centers of human self-realization. Thus, a century ago, Reclus had already presciently announced an intensifying crisis of the city and diagnosed this crisis as only a symptom of the larger crisis of society. He notes the multitude of "vices," such as poverty, crime, and ecological degradation that infest the modern city. He predicts that the continuing centralization of population in urban areas, with all its attendant problems, will continue to accelerate. Indeed, he foresees the rise of the gigantic megalopolis, and predicts that urban areas of ten and twenty million inhabitants will be "a normal phenomenon of social life" in the future.[59]

Reclus' comprehensive, critically holistic approach to urbanism made him an early critic of "urban renewal" schemes, which he attacks as based on a superficial view of urban questions. He points out that poor housing and bad health conditions are merely displaced when they are addressed through a problematic of renewing a certain delimited area rather than one of renewing the larger society and its natural environment. He notes pointedly that "in a society in which people cannot depend on having enough bread to eat, in which the poor and even the starving make up a large part of the population of every large city, it is no more than a halfway measure to transform unhealthy neighborhoods if the unfortunate people who previously inhabited them find themselves thrown out of their former hovels only to go in search of new ones in the suburbs, merely moving the poisonous emanations a certain distance away."[60] Most urbanists have yet to come to grips with this inescapable problem that was quite clearly diagnosed by Reclus at the dawn of the twentieth century.

For Reclus, cities cannot be fundamentally renewed so long as they remain an integral part of a corrupt and oppressive society. He admires creative and energetic attempts to improve the city and points to them as inspiring examples of what can be achieved in the future. However, he believes that true renewal can only result from reclaiming the city's rich heritage of personal freedom and vibrant local community that extended across a long history from the Greek polis down to the revolutionary democratic communities of the modern period. It is only when cities become expressions of the collective self-realization of all the citizens that they can possibly be renewed and regenerated. Only then can they become "perfectly healthy and beautiful organic bodies."[61] In short, the city can attain freedom, justice, beauty, and cooperation only when the social revolution achieves these goals for society as a whole.

Although the concept of mutual aid has been closely associated with the political philosophy of Kropotkin, Reclus deserves recognition for making this concept central to anarchist social theory at about the same time. Reclus declares mutual aid to be "the principal agent of human progress."[62] Like other revolutionary theorists, Reclus sees history as a struggle between the powerful and the masses who are oppressed and exploited. As a theorist of mutual aid, he also sees this struggle as a conflict between those who look upon society as an arena of competition in which some inevitably triumph over others, and those who envision a society in which social solidarity and cooperation prevail. He attacks the Social Darwinist idea that "the fittest," in the sense of the most effective individual competitors, must always triumph. He sees this theory as no more than an ideology aimed at legitimating the dominance of those with economic power.

Reclus contends that those who join together in the liberatory struggle to create a cooperative community will ultimately show themselves to be both the fittest and the most powerful. This, he thinks, will be demonstrated when their solidarity, combined with their superior numbers, allows them to overturn the prevailing economic and political system. "The law of the strongest will not always benefit the industrial monopoly," he predicts, for "the day is coming when might will be at the service of right."[63] The defenders of the status quo proclaim the eternal rule of "the law of the blind and brutal struggle for existence," but it will be succeeded by another law, that of "the grouping of weak individualities into organisms more and more developed, learning to defend themselves against the enemy forces, to recognize the resources of their environments, even to create new ones."[64] Thus mutual aid, allied with human intelligence, will once again show itself to be a force for social evolution.

Such creative self-organization was beginning to take place in Reclus' time within the cooperative movement. Both Elisée and his brother Elie participated very actively in that movement, helping to establish the first cooperative in Paris and collaborating to produce the journal *La Coopération*. However, Reclus' views began to change markedly after several disappointing experiences with cooperative efforts, and as he began to ally himself more closely with Bakunin and to participate actively in anarchist revolutionary organizations and the First International. He came to see a preoccupation with the creation of worker self-managed cooperatives and intentional communities as a diversion from the more crucial struggle against capitalism and the state. "As for us anarchists," he concludes, "never will we separate ourselves from the world to build a little church, hidden in some vast wilderness."[65]

Reclus' negative judgments concerning the cooperative movement and communal experiments seems to contradict his desire to transform human values and relationships in order to "make ready for the day" in which the new society will be achieved. Despite his strong emphasis on the importance of a dialectic between evolution and revolution in the process of social transformation, his ideas still retain some elements of the fetishism of revolution that was long endemic to the Marxist and anarchist left. His belief in evolutionary change as a precondition for revolutionary transformation was applied consistently to many areas, including personal life, educational efforts to spread progressive ideas, and the creation of anarchist organizations. On the other hand, he seemed to have much more limited faith in the development of counter-institutions that would put into practice his libertarian and communitarian values before the social revolution. The forms of organization that he promoted were primarily oppositional ones, such as revolutionary unions, revolutionary political groups, and radical political alliances. He did not seem to grasp fully the importance of developing a "pre-revolutionary" practice of libertarian and communitarian social life in areas like production, consumption, and cooperative living that would prepare the way for a more thorough transformation of society.

Nevertheless, one cannot deny the strength of Reclus' critique of cooperative experiments. He incisively points out the danger of economic cooperatives that are divorced from a larger movement for social emancipation. "One tells oneself that it is especially important to succeed in an undertaking that involves the collective honor of a great number of friends, and one gradually allows oneself to be drawn into the petty practices of conventional business. The person who had resolved to change the world has changed into nothing more than a simple grocer."[66] This astute diagnosis has been verified in a multitude of cases since Reclus' time and points out a danger inherent in many strategies for reform. Just as those who enter the dominant political system with radical goals often end by sacrificing those goals for the sake of success within that system, so those who participate in the dominant economic system with far-reaching goals often sacrifice their radicalism for the sake of success in that sphere.

It should also be noted that even as Reclus came to stress the greater importance of other forms of social transformation, he continued to see value in cooperative experiments. While they are, in his view, incapable of thoroughly changing present-day society, they are a good source of experience in the practice of the mutual aid that will form the basis of the future cooperative society. He believes that "studious and sincere anarchists" can learn much from those cooperatives that have "joined with one another to form ever larger entities in such a way as to encompass the most diverse functions, such as those of industry, transportation, agriculture, science, art, and entertainment," thereby developing a "scientific practice of mutual aid."[67] But despite these contributions, he believes that it is impossible for them gradually to expand and peacefully replace the existing system. He therefore concludes that rather than working in cooperatives, anarchists can better spend their time organizing a revolutionary movement that will seize power and quickly apply the principles and lessons of cooperation to the new social order.

Reclus subjects intentional communities to the same critique as he applies to other forms of cooperative endeavor. He contends that while these social experiments may indeed be remarkable achievements in themselves, they present no serious challenge to the dominant system of power. At best, they create a sphere of relative freedom on the fringes of that system without threatening it; at worst, they merely function as a part of the system and help stabilize it. He comments that intentional communities in the United States often succeeded materially, "only to allow themselves to be once again reabsorbed by the environment of all-powerful capitalism."[68]

This criticism of the "utopian" nature of intentional communities seems in some ways to be rather shortsighted. According to Reclus' own principles, an indefinite period of evolutionary change must precede the qualitative, revolutionary transformation of society. The potential of communitarian experiments had certainly not been exhausted in his time, and neither has it been today. In fact, it is quite clear that oppositional movements have directed only a minute fraction of their efforts into such experiments, while enormous efforts have gone into electoral activity, labor struggles, and revolutionary organization, none of which has effected the kind of transformation envisioned by Reclus. Thus, his argument against communalism can be leveled at all strategies for social change (including his own) that have not succeeded in overturning the prevailing system.

But if some of Reclus' arguments regarding intentional communities and cooperatives seem facile, some of his points are very well taken. He notes that some communitarians are "utopians" in the pejorative sense, in that they do not have a good grasp of the obstacles—especially the internal, psychological ones—that stand in the way of cooperation: "The persons who come together in order to form one of these societies with new ideals are themselves by no means completely rid of prejudices, old practices, and deeply rooted atavisms; they have not yet 'shed the old man.'"[69] Although the members of the community may physically leave the old society behind, they carry with them traces of the institutions that formed their character. Thus, "in the 'anarchist' or 'harmonist' microcosm they have created, they must always struggle against the dissociative and disruptive forces produced by habits, customs, the ever-powerful bonds of family, tempting advice from friends, the return of worldly ambitions, the need for adventure, and the obsession with change."[70] This is a very perceptive analysis. The problems Reclus points out are not, however, unique to cooperative or communal experiments. A similar critique can usefully be applied to any attempt to create new social relationships, including the revolutionary organizations and postrevolutionary institutions that Reclus himself advocated.

It should be noted that in his letters, Reclus sometimes expressed a much greater enthusiasm for intentional communities than is evident in his published writings. In a letter of 1902, he discusses his visit to an intentional community, the "International Brotherhood" community in Blaricum, Holland. He comments:

> What brave souls! With what courage they devote themselves to their work! With what nobility of language they discuss questions related to morality and humanity! How happy one feels in their good company! I took from them one of the most lasting impressions of my life. I felt myself truly to be among my brothers and your brothers, part of our great family.... Are these people "born again," to use your language? I believe so; I am confident of it. And if they are not born again, the zeal they exhibit today, their ardent desire for justice, will certainly have an influence on the imminent birth of those who will complete their work.[71]

In a letter of the next year he mentions several other intentional communities of varying tendencies and judges that despite their frequent difficulties in surviving in hostile environments, they "always have great importance in raising the level of morality around them."[72]

Although Reclus had mixed feelings concerning the potential of intentional communities and cooperative enterprises, he had a strong belief in the importance of the personal realm to the process of social transformation. As early as 1859, when he was still in his twenties, he writes to his sister Louise: "Let us found little republics within ourselves and around ourselves. Gradually these isolated groups will come together like scattered crystals and form the great Republic."[73] Later, in 1895, he elaborates on the nature of these "republics." The anarchist, he says, should "work to free himself personally from all preconceived or imposed ideas, and gradually gather around himself friends who live and act in the same way. It is step by step, through small, loving, and intelligent associations, that the great fraternal society will be formed."[74] One finds in this idea of small republics of everyday life the essence of what became widely known in later radical theory and practice as the "affinity group." Reclus made extraordinary efforts in his own personal life to apply his principles of mutual aid, freedom, and egalitarianism in this way. Some other prominent anarchists (most notably, Bakunin) preached equality and cooperation while often engaging in self-promotion, manipulation, and cabal. Reclus, on the other hand, sought consistency between his ideals and his practice. It was important to him that his own circle of family, friends, and coworkers constituted a small cell in the emerging organic community of freedom that he heralded in his theoretical writings.[75]

A final area in which Reclus made a strong positive conception to anarchist thought is the sphere of education. It is surprising that he has not been more widely recognized as one of the most important figures in the history of libertarian education, for many concepts often associated with Bakunin, Tolstoy, Ferrer, and other libertarian educational theorists were also proposed, and developed with at least as much originality, in his writings.[76] Reclus' conception of education focuses on the ideal of the free self-realization of the child. His ideas are in some ways reminiscent of Rousseau and also prefigure Montessori, Dewey, and other later reformers. He sees the primary objective of education as being to "help the child develop in conformity with the logic of its own nature. There is no need for any goal other than drawing forth in the young intellect that which it already possessed in an unconscious form, and to assist religiously with the interior labors of that intellect, without any hurry, and without drawing premature conclusions."[77] He sees this process of creating the conditions conducive to such natural unfolding as involving more than merely the intellect. Here, as in other areas, Reclus' approach is dialectical and holistic. He recommends that practical, physical endeavors always be combined with intellectual ones and stresses the fact that education must involve both the body and the mind. He says that if both intellectual abilities and "skill and muscular energy" are given due attention, there will be a "natural balance of power" in the developing human being.[78] The child's development also requires adequate motivation, so his or her interests, imagination, and "passion" must be encouraged. Fortunately, Reclus says, the educator has a powerful ally in pursuing this end—play. In his words, "free amusement is one of the great educators of man."[79]

Reclus' critically holistic and libertarian approach to learning demands a restructuring of education in accord with the child's stages of development. He believes that the practical faculties should be developed first, through more active and experiential forms of learning, combined with the use of stories and forms of play that develop the imagination. Throughout the educational experience "the direct study of nature and the consideration of its phenomena should become one of the principal elements."[80] For the young child, such study should avoid the mere assimilation of information and focus rather on the child's concrete experience of his or her natural surroundings. In Reclus' opinion, logic, science, and all fields stressing abstraction and generalities can be approached later with more success. He fears that an education that begins with the abstract will "deflower the imagination" and render the child "skeptical and blasé."[81] The direct experience of the natural world is the great educational resource not only for the instruction of children but for education in general. In the same spirit as Thomas Berry, who states that "the natural world itself" is "the primary educator," Reclus asserts that "the true school should be free nature, including not only the beautiful landscapes that one contemplates, and the laws that one studies in the field, but also all the obstacles that one must learn to overcome."[82]

Reclus argues against any kind of coercive or authoritarian methods of education. He contends that such means are entirely unnecessary if the needs, capacities, and interests of the child are considered carefully and subject matter is introduced at the appropriate time. When attempts to impose premature or unsuitable studies on a child result in failure, it is often concluded that coercion, rather than better education, is what is needed. In many cases, coercive methods are imposed before more effective noncoercive ones are even tried.

Reclus contends that a powerful noncoercive instrument in the educational process comprises the personal qualities of a skilled teacher. He notes that the capable educator has a kind of natural authority over the young child based on "greater size and power, age, intelligence, scientific knowledge, moral dignity and life experience."[83] This "authority of competence" (as it is often called in anarchist theory), combined with the natural activity and curiosity of the child, offers more than adequate stimuli to learning. Moreover, Reclus recommends a kind of Socratic method of helping draw out what is implicit within the student, combined with the Aristotelian assumption that human beings have a natural desire for knowledge: "The child wants to know and the educator wants to teach, that is, to show the child that he already knows unconsciously, and needs only to pay attention to things in order to know consciously."[84] In this sense, the educator only assists the student in the free realization of his or her intellectual potentialities.

Reclus stresses heavily the social dimension of education. He observes that the nature of the educational experience has a powerful influence on a child's development as a social being and on his or her future capacity to participate cooperatively in the life of the community. The character of the learning group is therefore crucial. On the one hand, it must be large enough to create a spirit of collective endeavor as the students pursue their interests. They will thereby learn lessons in cooperation that they can later apply in their personal lives, their work, and their communities. On the other hand, the group must be small enough for a close relationship to exist between all the students and the instructor. Such a group will form "a veritable family for the joys of work and play."[85] The instructor will be "both a father and a brother," having the natural authority that comes from age and competence, but will always consider the students' needs, interests, and developing autonomy.[86] In effect, the libertarian educational group will combine certain qualities of the affinity group with others of the cooperative workplace. It will therefore help the child develop into a person who can participate successfully in both these realms.

Reclus' analysis of educational methodology stresses, in a quite Deweyan spirit, the distinction between education arising out of direct experience and engaged understanding, and education based on abstract dogmatism and sterile abstractions. In a letter of 1881 he proclaims: "I hate textbooks. Nothing is more detrimental to the intellectual health and the morale of the students. They present science to the student as something ready-made, finished, signed and approved, made almost into a religion and on the way to becoming a superstition. It's a diet that is dead and that kills."[87] He adds that "for science to come alive, it is necessary that [the student] live it himself, that he create it, and so to speak, renew it constantly."[88]

It must be added that for Reclus, formal education is only one aspect of a larger process of libertarian education—education for social self-realization—within society as a whole. He explains that "the ideal of the anarchists is not to eliminate the school, but rather to enlarge it, to make society itself into a great body for mutual instruction, where all will be at once pupils and professors, where each child, after having received 'the basics' in primary education will learn how to develop himself integrally, according to his own intellectual capacities, in the existence that he has chosen for himself."[89]

Reclus' ideas concerning formal education are thus an application of his more general theory of the development of liberatory consciousness and practice within society. They reflect what has often been seen, with good reason, as one of the great strengths of his political thought: his analysis of the close relation between revolutionary change at the level of social institutions and prior evolutionary change at the level of personal life, values, and social practices. In view of the perceptiveness and originality of much of his analysis of social evolution and revolution, Reclus deserves greater recognition than he has usually been given, not only as one of the major figures in the history of anarchist thought but as a significant figure in modern European social and political theory in general.

6

The Critique of Domination

Reclus was always an anarchist by temperament, and his libertarian ideas began to develop early in his life. However, his anarchism became increasingly deeper and more consistent as his social analysis expanded into a detailed critique of all forms of domination. As has been shown, Reclus was unusual in questioning the conception of human domination of nature that was accepted not only by mainstream thinkers but also by most critics of the prevailing order. He is also distinctive for the comprehensiveness of his critique of domination within human society. His analysis of various institutional forms of social domination constitutes one the most far-reaching accounts in classical radical thought, and his position prefigures in many ways the more widely recognized achievements of later critical theory.

One of the most extensively developed aspects of Reclus critique is his devastating attack on all forms of the state, including what he saw as the ideological fiction of the "representative" state. Although as late as 1871 he was willing to offer himself as a candidate for the National Assembly, he soon came to oppose the parliamentary system entirely, and for the remainder of his life he rejected voting in national elections, even in the proverbial case of "the lesser of evils." In his opinion, all those who seek to exercise power in a centralized nation-state render themselves vulnerable to absorption into that system of domination. He says that aspiring officeholders are "raised above the crowd, whom they soon learn to despise," so that they "end by considering themselves essentially superior beings; solicited by ambition in a thousand forms, by vanity, greed and caprice," and "are all the more easily corrupted."[1] He observes that their corruption is encouraged by a "rabble of interested flatterers" who are "ever on the watch to profit by their vices."[2]

Reclus' remarks on the processes by which elected officials are selected are often quite astute and describe accurately many features of what is called representative democracy. He notes that in order to gain support, an aspiring public official must please a variety of factions so that "ambitions inevitably emerge, and machinations, extravagant promises, and lies have free rein. Moreover, it is certainly not the most honest candidate who has the best chance of winning."[3] In theory, the legislator should be a specialist in every area in order to make decisions for everyone on every subject imaginable. But in reality the candidates do not possess such capacities any more than do the members of the electorate. In practice, what is required of them is only that they be experts at getting elected, and "no particular ability recommends the candidate to the voters."[4] Entirely irrelevant or arbitrary qualities become essential to electoral victory, and "the winner may owe his success to a certain provincial popularity, his good-natured qualities, his oratorical skills, or his organizational talents, but frequently he is also indebted to his wealth, his family connections, or even the terror that he can inspire as a great industrialist or large property owner."[5]

The products of this degraded system, Reclus contends, are a collection of mediocrities who have no conception of the common good. The successful politician "will be a man of the party; he will be asked neither to involve himself in public works, nor to facilitate human relations, but rather to fight against one faction or another."[6] The greatest danger is not the incompetence of the legislature but the fact that it is "inferior in moral qualities, since it is dominated by professional politicians."[7] In Reclus' opinion, these falsely denominated "representatives of the people" will certainly make far worse decisions for the populace than its individual members would have made for themselves without going through the trouble of holding elections.

Reclus contends that after the so-called representatives are elected, they are free to move ever farther from any kind of effective control by the people. Knowing that there is no real accountability between elections, and "well aware that he can now commit crimes with impunity, the elected official finds himself immediately exposed to all sorts of seductions on behalf of the ruling classes."[8] The legislators find themselves in a world of power and wealth that is quite alien to the lives of those who elected them. The power of this milieu overcomes any scruples that may have stood in the way of identification with the political elite, as "the newcomer is initiated into the legislative traditions under the leadership of the veteran parliamentarians, adopts the *esprit de corps*, and is solicited by big industry, high officials, and above all, international finance."[9] As is typical of all modes of socialization, the immediate social environment has an overwhelmingly powerful effect on the individual.

Reclus' attack on electoralism is directed especially against the parliamentary socialists. He comments that "it would be absurd on our part to hold a grudge against the socialist leaders who, finding themselves caught up in the electoral machine, end up being gradually transformed into nothing more than bourgeois with liberal ideas. They have placed themselves in determinate conditions that in turn determine them."[10] But while he (perhaps disingenuously) dismisses the idea of "holding a grudge against" such socialists, he certainly thinks they should be exposed as traitors to the cause of freedom and justice. Since they use the rhetoric of justice and equality and claim to act on behalf of the masses, they become powerful impediments to the popular acceptance of libertarian and revolutionary ideas. They sap the people's revolutionary energies by diverting them into ineffectual and indeed counterproductive strategies. Reclus would no doubt direct similar criticisms at today's "progressive" and social democratic politicians who argue that society can be made just and free if only their faction is elected to national office in sufficient numbers to guide the policies of the nation-state and mitigate somewhat the worst excesses of the capitalist system.

According to Reclus, such politicians corrupt themselves and become shameless hypocrites in seeking electoral success and high position rather than devoting all their efforts to achieving the social emancipation to which they profess allegiance. Once again (as on the issues of cooperatives and intentional communities), he raises the question of a discontinuity between means and ends. He argues that means that are justified in the name of quite laudable ends soon come to exert such a powerful influence that the ends become no more than ideological alibis for helping to perpetuate the system of domination and to legitimate the position of the nominal opposition within that system. In pursuit of electoral success, "the socialist candidate readily flatters the tastes, the inclinations, or even the prejudices of his electorate. He blithely ignores disagreements, disputes, and grudges, and for a while becomes the friend, or at least the ally, of those with whom only a short time ago he had exchanged invectives."[11] If these politicians succeed in their quest for office, the disintegration of their character only accelerates. "Their very spirit undergoes a pervasive transformation," and they finally end up as "experts at exchanging smiles, handshakes, and favors."[12]

Reclus contends that the achievement of electoral success, though always greeted with euphoria, may actually be a major setback for any growing movement for social change, no matter how just and moral its professed ends may be. While it would seem that such a movement has gained a great victory if it can enact its demands into law, "it is possible that the result will be precisely the opposite. While it is true that any charter or laws that are agreed to by the insurgents may sanction the liberty that has been won, it is also true that they will limit it, and therein lies the danger. They determine the precise limit at which the victors must stop, and this inevitably becomes the point of departure for a retreat."[13] Within a vital social movement, ideas and practices develop and evolve among the community of those committed to common ends. The shift to the parliamentary realm (and often even merely to the electoral one) transforms this dynamic process into a debilitating struggle to salvage some elements of the ideal in the form of a viable "program" or "platform." This critique perhaps overstates the extent to which all such efforts to embody far-ranging ideals in practical political programs must founder on the shoals of political pragmatism. Yet the examples multiply of "progressive" (whether liberal, social democratic, or green) social movements that have envisioned vast social transformations, only to become practically ineffectual and theoretically insipid in the face of their own electoral and legislative successes.

If Reclus is harsh in his criticism of legislative power, he is no less scathing in his attack on bureaucracy and administration. He observes that injustice is woven into the fabric of existing society and that as long as the system of power remains fundamentally intact, any large institution must adapt itself to the prevailing conditions. He has a disturbing yet powerful message for many liberals, social democrats, and others of the reformist left who have continued to believe that an improved state bureaucracy is a promising agent for the rectification of social injustices. "As soon as an [administrative] institution is established, even if it should be only to combat flagrant abuses, it creates them anew through its very existence. It has to adapt to its bad environment, and in order to function, it must do so in a pathological way. Whereas the creators of the institution follow only noble ideals, the employees that they appoint must consider above all their remuneration and the continuation of their employment."[14] Far from constituting, as Hegel claimed, a "universal class," the bureaucracy is rather a powerful particularistic interest.

Furthermore, Reclus warns, bureaucracy is hopelessly inefficient. While bureaucratic rationality claims to maximize efficiency, it in fact does the reverse, because "first, it impedes individual initiative in every way and even prevents its emergence; second, it delays, halts, and immobilizes the works that are entrusted to it."[15] Reclus' critique of bureaucracy is of interest in part because many of his points sound so much like the antibureaucratic discourse of the neoliberal right in many countries today. But Reclus would argue that the contemporary left succumbs to a fallacious and disastrous logic when it deduces from the true premise that big business acts in ways detrimental to society the false conclusion that state bureaucracies will act in ways that are truly beneficial to the populations whose welfare is entrusted to them. He would contend that this logic will seldom seem convincing to those who have the most immediate experience of the administrative state, the actual "clients" (he would say "victims") of state bureaucracies.

Reclus argues that the bureaucracy creates its own forms of social irresponsibility and greatly reduces efficiency at the same time. Despite his loathing for the capitalist industrial system, Reclus finds the bureaucratic world to be in some ways even worse. While human values are trampled on by industry, at least its economic rationality produces an effort to reduce waste and increase productivity. On the other hand, "the administrative hierarchy does its utmost to multiply the number of employees and subordinates, directors, auditors, and inspectors. Work becomes so complicated as to be impossible," and "everything becomes a pretext for a delay or a reprimand."[16] In this respect, there is even less redeeming social value in state bureaucratic organization.

Reclus sees another evil of bureaucracy in the loss of responsibility that results from its complex, mazelike network of power, in which accountability becomes impossible. In such a system, "minor officials exercise their power more absolutely than persons of high rank, who are by their very importance constrained by a certain propriety," and "the petty official need not have the slightest fear of being held responsible in this way so long as he is shielded by a powerful boss."[17] The egoistic, dominating personality fostered by authoritarian institutions gains a multitude of outlets in the bureaucratic labyrinth: "The uncouth can give free rein to crass behavior, the violent lash out as they please, and the cruel enjoy torturing at their leisure."[18]

Reclus thus developed a quite challenging critique of bureaucracy over a century ago, pointing out evils that have become even more evident in the massively bureaucratic states that have developed since his time. It might seem ironic that the mainstream left has allowed the right to monopolize antibureaucratic rhetoric for its own benefit, even as military and corporate bureaucracies hostile to the presumed goals of the left have proliferated. Reclus would argue that this lapse has continued to exist for very good ideological reasons. He would point out that an incisive critique of bureaucratic abuses would necessarily lead to a critique of statism, party politics, and all the other related hierarchical institutions that oppositional political movements and parties have themselves done so much to perpetuate and in which they continue to have a vested interest.

Another area in which Reclus' critique of the state is particularly acute is in his discussion of patriotism and the ideology of nationalism. His dissection of the extremes of patriotic folly prefigures later analysis of the psychology (and pathology) of nationalism by figures such as Randolph Bourne and Wilhelm Reich. Although individuals, he notes, may sometimes escape a prevailing insular mentality, masses of people tend to remain in the grip of a "primitive morality of force" that can be aroused whenever their leaders find an enemy against whom they can direct malignant passions and murderous fantasies.

Reclus is quite eloquent in his description of the collective insanity that periodically breaks out under the influence of nationalist leaders and manipulative politicians. Once it is deceived into conflict, a nation unites in patriotic hatred and then "delights in ravishing, killing, and then singing of victory over the sprawling corpses. It glories in all the evil that its ancestors have inflicted on other peoples. It gets carried away, and wildly celebrates in verse, in prose, and in triumphant depictions, all the abominations that its own people have committed in foreign lands. It even solemnly invites its God to take part in the general intoxication."[19] What Reclus describes, though he never specifically mentions the term, is a process in which all the forces of the *social imagination* are enlisted on behalf of the state and its atrocities.

This mechanism is one of the most powerful means of control available to "the masters of the people." Social antagonisms are calmed and oppositional forces abated by "transforming all the energies of a nation into a rage against the foreigner."[20] This recourse is readily available since the state and its rulers labor tirelessly to rewrite history as a chronicle of offenses against the nation by malevolent foreign powers, with episodes of injury and victimization alternating with those of triumph and revenge. The fundamentally oppositional and antagonistic nature of the system of nation-states greatly facilitates this task. Pretexts for turning neighbors into enemies "are easy to find, since the interests of states remain different and in conflict through the very fact of their separation into distinct artificial organisms."[21] In addition, given the vicissitudes of international relations, there is usually a long history of conflict at the disposal of the rulers for purposes of inflaming the public. They can make use of "the memories of actual wrongs, massacres, and crimes of all sorts committed in former wars. The call for revenge still resounds, and when a new war will have passed like the terrible flames of a fire devouring everything in its path, it will also leave the memory of hatred and will serve as leaven for future conflicts."[22]

The oppositional nature of the state dictates not only conflict but also conquest. By Reclus' time the age-old process of state expansionism had evolved into the system of global imperialism in which the more powerful European nation-states had succeeded in subjecting most of the surface of the earth to colonization. This process of global transformation was of great interest to Reclus both as a social geographer and as a political theorist. Béatrice Giblin notes that his views on colonization are ambiguous because of his distinction between "colonies of exploitation," based on the domination of conquered peoples, and "colonies of population," which, in his view, have contributed to progress through the spread of constructive human activity over the face of the earth.[23] Giblin correctly points out that he "had a dialectical vision of the phenomenon of colonialism," according to which "he denounces the negative effects—such as the plunder of the economy, the destructuring of indigenous cultures, the increase in famines, etc.," while also recognizing such positive consequences as "the spread of education to a greater proportion of the population, the progressive disappearance of certain 'barbaric' customs such as infanticide of female offspring, improvement in health conditions, etc."[24] His assessment of colonialism thus coincides with his general view of social phenomena, which requires a careful analysis of both progressive and regressive moments.

It would be incorrect to see in this dialectical analysis any sort of apology for colonization. Reclus does not seek to explain away the evils of history in any quasi-Hegelian defense of their necessity in some vast scheme of world history. Rather, he asserts that freedom and solidarity must always be defended and that progress must be pursued only through the most just and liberatory means possible. He therefore vehemently opposes the spread of imperial state power, whatever good might be discovered amidst the evils of conquest and oppression. For example, his condemnation of the French colonization of Algeria—despite the implantation of widespread "colonies of population" there—is scathing. He notes that the military conquerors "were interested much less in the fate of the conquered populations than in plying their trade, and he saw little in Algeria beyond a vast field for training troops in all sorts of military exercises," so that as a consequence "the so-called military spirit was formed—a spirit hostile to all free thought, individual initiative, and peaceful, spontaneous progress."[25] He condemns colonial exploitation elsewhere on similar grounds, sometimes noting the correspondence between the destruction of indigenous cultures and the assault on the integrity of nature as European power spread across the earth.

While Reclus finds certain evils, such as conflict and conquest, to be inherent in the very form of the nation-state, he does not treat "the state" as a monolithic institution. He recognizes that states have diverse histories and compositions, and that they play varying roles in the course of history. He is even willing to concede that they have both progressive and regressive aspects (perhaps a surprising concession for as thoroughgoing an anarchist as Reclus). In his consideration of modern nation-states, he finds Russia and the United States to be of particular importance in the future course of history. He sees these two societies as the major modern paradigms of social organization. The United States is "a republic, the leader of other republics," while Russia represents "conservative principles and the old despotism."[26] Indeed, he discovers in Russia the seeds of an Oriental despotism that constitutes a major threat to European society. He warns of "the poison of a traditional, atavistic servitude that easily spreads through the veins of the European: the Oriental conception of the necessity for a strong government," and suggests that there is no lack of "base souls, happy to renounce themselves and to obey."[27] In particular, he fears that European authoritarianism would be promoted by the example of a modern Russian despotism that carried on the tradition of Genghis Khan and Ivan the Terrible. It takes little imagination to combine these remarks on Oriental despotism with Reclus' attacks on authoritarian socialism and to discover a powerful implicit critique of Stalinism and the future Soviet "communist" state.[28]

Reclus also warns against more subtle developments in the direction of despotism. He recognizes that with the spread of illusory "representative democracy" the state begins in some ways to gain a firmer hold over the populace. "On the one hand, the ambition to govern becomes widespread, even universal, so that the natural tendency of the ordinary citizen is to participate in the management of public affairs. Millions of men feel solidarity in the maintenance of the state, which is their property, their affair."[29] On the other hand, the state is strengthened by the spread of an equally superficial social democracy, as a multitude of people become dependent on it for "small entitlements to income."[30] For Reclus, no less than for Friedrich Hayek, the social-democratic state is certainly a "road to serfdom" (though Reclus would contend that Hayek and other conservative critics of the state merely take detours to the same destination).

But in spite of his fears of regression to old forms of despotism and evolution toward new ones, Reclus holds out hope that the despotic state might on the whole be moving into a period of decline. He notes that even as the public begins to find itself on much more intimate terms with the state, true to the cliché, familiarity begins to breed contempt. The unmasking of the state's cynical exercise of power begins, and the manipulative and self-interested nature of its actions becomes increasingly clear. Reclus is rather prophetic in describing this unmasking process, which might be called the "disenchantment of the state." While we have yet to see whether it proceeds to what he sees as its logical conclusion, his analysis of the phenomenon is brilliantly prescient.

He observes that as the populace becomes more involved, albeit superficially, in the affairs of state, "this banal government, being all too well understood, no longer dominates the multitudes through the impression of terrifying majesty that once belonged to masters who were all but invisible and who only appeared before the public surrounded by judges, attendants, and executioners."[31] While the popularization of the state superficially seems to reinforce its power, the disenchanted state loses its capacity to "inspire mysterious and sacred fear" and finally reaches the point at which "it even provokes laughter and contempt."[32] He suggests that historians will have to study satire and caricature to understand adequately the fate of the state and government beginning with the second half of the nineteenth century.

Reclus describes this process in dialectical terms: "The state perishes and is neutralized through its very dissemination. Just when all possess it, it has virtually ceased to exist, and is no more than a shadow of itself."[33] The transformation of the modern state illustrates the classically Hegelian dialectical principle that when a phenomenon reaches the limit of its self-development it begins to destroy itself, though its history is preserved through its embodiment in succeeding phenomena. "Institutions thus disappear at the moment when they seem to triumph. The state has branched out everywhere; however, an opposing force also appears everywhere. While it was once considered inconsequential and was unaware of itself, it is constantly growing and henceforth will be conscious of the work that it has to accomplish."[34] The opposing force is, of course, the movement for human liberation. Needless to say, Reclus' optimism concerning its efficacy, and even its growing self-consciousness, has yet to be borne out in history. And to the extent that the legitimacy of the state has indeed eroded, its place has often been taken by other modes of domination, including the form that is given the most extensive attention in Reclus' own analysis—economic exploitation.

Although Reclus launches a stinging attack on the state and bureaucracy, it is economic power that is the object of his most far-reaching critique. In his view, capital is the supreme power in modern society and the major obstacle to social emancipation. He therefore presents an extensive analysis of the evolution of forms of property, the domination of society by economic power, and the destructive effects of the economization of society and its values. His reflections on the subject make an important contribution to an anarchist theory of property. In part, he further develops such conceptions as Godwin's idea of entitlement based on need and Proudhon's distinction between exploitative forms of property, which the latter defined as "theft," and property as personal possession, which he saw as a form of "freedom."

According to Reclus, there are ancient forms of appropriation that preceded "property" as we now conceive of it. Early societies linked possession to use and had no conception of individual or group ownership. Even collective property, he says, is "a limitation of the primitive right to labor belonging to all."[35] The most ancient forms seem to come closest to what Reclus proposes for the future, which is a kind of distribution according to need, or, as he would put it, distribution based on solidarity with others and with the community. In his view, the earliest forms were succeeded by a system of possession of property by the community as a whole. Although he sees a regressive aspect in this change since it introduced ideas of possession that had potentially antisocial implications, he argues that many of the virtues of the more ancient system were preserved. Indeed, he praises lavishly what remains of the tradition of communal property, for it offers a sphere of resistance to the domination of capitalist property relations, presents a point of reference with which to show the brutality of the present economic order, and points toward a future system of cooperative production.

Critics have sometimes contended that anarchist thought, and classical anarchist theory in particular, has emphasized opposition to the state to the point of neglecting the real hegemony of economic power. This interpretation arises, perhaps, from a simplistic and overdrawn distinction between the anarchist focus on political domination and the Marxist focus on economic exploitation. While there is abundant evidence against such a thesis throughout the history of anarchist thought, Reclus' analysis refutes it in a particularly conclusive manner. In his view, "one overriding fact dominates all of modern civilization, the fact that the property of a single person can increase indefinitely, and even, by virtue of almost universal consent, encompass the entire world."[36] He observes that the ability of capital to transgress all boundaries of state and nationality gives it a great advantage over political power. "The power of kings and emperors has limits, but that of wealth has none at all. The dollar is the master of masters."[37]

In analyzing the destructive aspects of capital, Reclus considers arguments that concentration of property has fostered economic and technological progress, and that economies of scale have increased productivity. He holds that if one looks at this concentration from the standpoint of social geography rather than from that of economic rationality, the results are seen to be detrimental to both human society and to nature. In his words, "the devouring of the surrounding land by the large estates is hardly less disastrous than fire and other devastations" since "it produces the same end result, which is the ruin not only of populations but also frequently of the land itself."[38] With stinging irony he notes that "intelligent large landholders can no doubt train excellent farm hands, and they will certainly have domestics of impeccable correctness," but they make no contribution to social progress since they produce "subjects" rather than "dignified equals."[39] Reclus the social geographer once again shows how the ecological and the social (in this case degradation of the land and degradation of character) are intimately interrelated and result from the same root causes.

It has been noted that Reclus saw Russia and the United States as the two emerging models for the next period of world history. While Russia is taken as the paradigm for statism and political domination, the United States serves as the model for economism and the power of capital. Nevertheless, Reclus is not so naïve as to reduce American society to its economic system in the manner of some superficial social critics, including some anarchists. He knew that society well and presents a subtle analysis of both its great achievements and promise as well as its tragic injustices and contradictions. He finds its republican institutions to have many admirable qualities, and he recognizes its capacity to exert a positive influence on societies that are still struggling to end political despotism. In addition, he praises it as the world's best example of "daring, initiative, and energy in labor."[40] He suggests, however, that much of the society's vast potential for social progress is negated by the dominance of its economistic values.

After living several years in the United States, he depicted in his early work *A Voyage to New Orleans* the baneful effects of commerce on the American character. He notes that while Europeans "obey traditions rather than humans" and are "governed by the dead more than by the living," in the United States "not a single superstition is attached to the past, or to the native soil, and the population, moving like the surface of a lake seeking its level, distributes itself entirely according to the laws of economics."[41] He observes that in such a society the quest for innovation leads to widespread destruction, as so much falls quickly into obsolescence. He notes that "in the young and growing republic, there are already as many ruins as in our old empires."[42] Writing before the Civil War, he diagnoses rather acutely the growing dominance of economic ideology in American society. "For the masses," he comments, "all feelings merge more and more with pecuniary interests."[43] He was greatly alarmed at the prospect that the "leader of republics" would lead other nations toward class domination and egoistic exploitation rather than freedom and solidarity, as it spread this economistic outlook throughout the world.

Indeed, Reclus foresees the coming ascendancy of a world economic system pervaded by a global culture of economistic values. This culture, he notes, is already entrenched in societies of the West. He remarks that "for the typical civilized European, or better yet, the North American, the essential thing is to train oneself to pursue monetary gain, with the goal of commanding others by means of the omnipotence of money. One's power increases in direct proportion to one's economic resources."[44] Such a system of values is dominant in countries with European cultures but is spreading across the globe. He notes its influence in "those countries of Asia that have developed in the direction of the ideal world of economics, and in all other parts of the world that are carried along by the example of

Europe and its all-powerful will."[45] It would not be surprising to Reclus to find that today countries of East Asia have been integrated into the core of the world system.

Reclus laments the fact that as this process of economization and Europeanization extends to an increasing number of traditional cultures, the communal traditions that might contribute to the development of a free communitarian society are destroyed: "The ancient forms of property that grant to each member of the community equal right to the use of the earth, water, air and fire are nothing more than archaic survivals in the process of rapid extinction."[46] Whereas Marx and many other classical radical theorists accepted the dissolution of all traditional institutions as the necessary cost of economic and technological progress, Reclus sees the processes of commodification and economic rationalization as destroying genuinely progressive features of traditional culture—features that might be developed in a liberatory direction.

In addition to diagnosing at a rather early date the destructive cultural effects of global economic imperialism, Reclus also warns of the dangers inherent in the more purely economic tendencies of that system. Describing the process of economic expansion, he observes that

> the theater expands, since it now embraces the whole of the land and seas. But the forces that struggled against one another in each particular state are precisely those that fight across the earth. In each country, capital seeks to subdue the workers. Similarly, on the level of the broadest world market, capital, which has grown enormously, disregards all the old borders and seeks to put the entire mass of producers to work on behalf of its profits, and to secure all the consumers in the world, savage and barbarian as well as civilized.[47]

He notes that in this globalizing economy the state acts increasingly on a transnational level to enforce the interests of economic power. He observes that already troops have been dispatched to foreign countries "by order of the stock exchange" and that "the unlimited power of capital and its international character are phenomena that are so well established that one may speak of the replacement of governments by banks in directing the undertaking and administration of war and peace."[48] Finally, he goes so far as to ask rhetorically whether these economic institutions that "already directly manage—albeit under an assumed name—billions of the budget, do not indirectly manage all the affairs of state."[49] He thus depicts modern capitalism, at a relatively early stage in its development, as an increasingly totalizing system of global economic imperialism.

Reclus' critique of economic domination is based above all on his acute sense of justice, on his deep compassion for those who suffer, and on his intense feeling of outrage at the subordination of some to the power of others. His concern for justice is expressed perhaps most strikingly in such popular works as his pamphlet *To My Brother the Peasant*, in which he juxtaposes a description of the enormous wealth of the infant born into the world of privilege and an account of the growing degradation of the lives of working people. Here and in some of his other more polemical works he expresses, often in eloquent terms, the righteous moral indignation that underlies his social analysis.

But despite his identification with the oppressed, he does not neglect the detrimental effects of injustice on those who seemingly benefit from exploitation. He believes deeply that a life of cooperation and mutual aid within a compassionate community is the most fulfilling existence for a human being. Conversely, he sees a life of privilege based on injustice as no more than an illusion of happiness and success. For Reclus, the human spirit is necessarily distorted when some prosper at the expense of others. Accordingly, "the present cruel state of inequality, in which some are overloaded with superfluous wealth while others are deprived even of hope, weighs like a bad conscience on the human soul, whether one is aware of it or not. It weighs most on the souls of the fortunate, whose joys are always poisoned by it."[50] Reclus' critique of economic inequality is fundamentally an ethical one that focuses on the fact that it destroys the human potential to achieve the good life, whether on the part of the oppressed or of the oppressors.

Reclus also does not overlook the many ways in which the drive for economic power destroys what is of value in the natural world and prevents it from flourishing. In his relatively early essay on "The Feeling for Nature in Modern Society," he describes the acquisitive drive to turn into a new source of profit everything in nature that can possibly be exploited economically. He notes that even the most exalted spiritual and aesthetic dimensions of nature are increasingly reduced to the level of economic values.

> Each natural curiosity, be it rock, grotto, waterfall, or the fissure of a glacier—everything, even the sound of an echo—can become individual property. The entrepreneurs lease waterfalls and enclose them with wooden fences to prevent non-paying travelers from gazing at the turbulent waters. Then, through a deluge of advertising, the light that plays about the scattering droplets and the puffs of wind unfurling curtains of mist are transformed into the resounding jingle of silver.[51]

Reclus' discussion of such economic exploitation exhibits the appeal to both moral and aesthetic sensibilities that so often marks his ethical critique.

Reclus was also an early and perceptive critic of the social regimentation and control that results from the development of industrial technology at the service of economic power. He deserves recognition for his prescient insights concerning technocracy in general and the dangers of the coming machine civilization. In his widely reprinted *To My Brother the Peasant*, he warns that "we are in an age of science and method, and our rulers, served by an army of chemists and professors, are preparing a social structure for you in which all will be regulated as in a factory. There, the machine controls everything, even men, who are simple cogs to be disposed of when they take it upon themselves to reason and to will."[52] In some ways, his critique is reminiscent of Marx's account of alienation and dehumanization under a capitalist division of labor. However, unlike his great contemporary, he did not see humanity's passage through "the steeling school of labor" as a historically necessary stage, and he did not accept the desirability of continuing the regimentation of labor until the automated industrial machine freed humanity from the necessity of toil.

In order to illustrate the nature of the process of mechanization, Reclus points to the example of industrialized farming in the American West. The criteria for the organization of production, he notes, are, first, the reduction of everything, including human beings, according to technical rationality, to quantifiable and manipulable resources, and, second, the efficient use of these resources according to the dictates of economic rationality, with minimal investment and maximum return. "Machines, horses, and men are used in the same manner: they are viewed as so much force to be quantified numerically, and they must be used most profitably for the employer, with the greatest productivity and the least expense possible."[53] The result is a system of regimentation and control in which "all of the workers' movements are regulated from the moment they leave the communal dormitories."[54]

Later, in *Man and the Earth*, he describes this system as a process of reducing workers "to the simple role of living cogs in the machine" who, after "repeating the same motions millions or even billions of times," finally have "but the appearance of life."[55] He thus outlined and criticized the principles of capitalist "scientific management," just as Frederick Taylor was beginning to introduce it into American industry. Of course, Adam Smith had described the process of mechanized labor, and Marx and others had criticized it long before. But Reclus' analysis goes farther than many others, for in addition to stressing the dehumanization, alienation, and immiseration of the workers, he also points out the emergence of systematic technological domination of society. His judgment of the developing system of technical rationality is astute: "Never did ancient slavery more methodically mold and shape human material to reduce it to being a tool."[56] In his view, modern "free" labor is subjected to a kind of domination and objectification that was inconceivable under any previous system of exploitation, no matter how brutal it may have been.

Bakunin is often thought to be distinctive among anarchist political theorists for his prophetic warning of the coming domination of the "new class"—that is, the rule of the new elite of scientific and technical intelligentsia. In Bakunin's words, their attainment of power would signal "the reign of *scientific intelligence*, the most aristocratic, despotic, arrogant and contemptuous of all regimes."[57] However, Reclus also stresses the dangers posed by the rise of this new class. He notes the example of those German scientists who served as "intellectual bodyguards for the imperial House of Hohenzollern"[58] and warns that although the technobureaucratic elite can function as an important tool of concentrated power, it also aspires to the direct exercise of such power: "If some scientists pride themselves in serving the master, there are others who aspire to become masters themselves."[59] He cites the principles of the Saint-Simonians and the Comptists as early examples of ideologies legitimating technocratic domination in the alleged interest of society as a whole. According to these schools, society must be managed "like a great factory directed by the discretion of engineers" and, more particularly, directed "precisely by the leaders of these new schools, who aim at infallibility."[60] In his view, this project of technocratic domination was advanced by authoritarians of his own time under the guise of "scientific socialism."

Reclus' reaction to the growth of scientific and technical knowledge and expertise was far from entirely negative. Here as elsewhere his view is a dialectical one, in which both positive and negative moments are identified and analyzed. He finds that despite its ruthless and destructive aspects, the developing scientific and technical division of labor has made a definite contribution to social progress. It has done so not only by increasing the wealth of society, as its defenders often claim, but also by fostering "the participation of an increasingly greater number of workers in the science of mechanics and all the associated areas of knowledge, including electronics, chemistry and metallurgy."[61] Much like Bakunin, Kropotkin, and other anarchist theorists, Reclus believes that society's goal should be to extend this process to create a "synthesis of intellectual and manual labor" in which "science becomes active."[62]

For Reclus, the distinctive quality of the division of labor and the ideal toward which it moves is not mere increase in production but above all the creation of "solidarity between all the functions that are divided from one another."[63] In his view, modern industry subverts this ideal by using the division of labor to increase its profit and, in pursuit of this goal, seeks "to separate the workers, isolate them one from another, and maintain its own power through the breaking up of opposing forces."[64] The challenge for a liberatory social movement, according to Reclus, is to take control of the system of production so that the growing solidarity of labor can be allowed to develop freely and take its place within a larger system of solidarity encompassing both humanity and nature.

Another form of domination that concerned Reclus very deeply throughout his life was racism. For Reclus, racism expresses a contemptuous, hierarchical ranking of human beings that conflicts with his sense of human solidarity, his belief in social equality, and his respect for the achievements of all cultures. This great social evil became a matter of intense personal interest to him when his stay in Louisiana gave him direct experience of a racist, slaveholding society. His marriage to a woman of mixed African and European ancestry intensified his personal involvement in the issue. While classical anarchist thinkers and radical theorists in general tended to focus their critique on the state and capitalism, Reclus always identifies racism as one of the most pernicious forms of oppression and domination. He holds that the most effective response to racism is the destruction of the social barriers that have been created to enforce the system of dominance and subordination. In his view, society is always strengthened by the creative diversification resulting from the interaction and blending of peoples and cultures. The social and biological intermingling of previously segregated races would therefore both eliminate the basis for racism and also contribute to the vitality of society.

Reclus was particularly interested in the conditions of black people in the United States, a topic he analyzed both before and after slavery. In his *Voyage to New Orleans*, he presents a moving depiction of an antebellum slave market. Reflecting on the atrocities of the auction block, he observes that "all the Negroes of Louisiana pass in turn on this fateful table: children who have just ended their seventh year and whom the law in its solicitude deems old enough to be separated from their mothers, young girls subjected to the stares of two thousand spectators and sold by the pound, mothers who come to see their children stolen from them, and who are obliged to remain cheerful under threat of the whip, and the elderly, who have already been auctioned off many times, and who have to appear one last time before these pale-faced men who despise them and jeer at their white hair."[65] His encounter, relatively early in his life, with racism in this most brutal form affected him profoundly and left him with an unusual sensitivity to the ethnic dimensions of domination.

Reclus was well aware of the fact that the abolition of slavery did not eliminate the system of racism and the exploitation of black people in America. He notes that after what was called emancipation, capitalist entrepreneurs found ways to exploit the freed labor power of former slaves at the lowest possible cost. The result was "slavery, minus the obligation to care for the children and the elderly."[66] New discriminatory laws and the biased enforcement of existing ones facilitated this exploitation by segregating blacks in living areas near plantations and workplaces, and by depriving them of the vote. In some areas, imprisonment for minor infractions was encouraged so that entrepreneurs could make use of forced prison labor. Some towns, consumed with "pure, brutal and instinctive hatred," merely expelled blacks and forbade their reentry.[67]

Reclus was unusual among social critics of his day in developing an extensive critique of American racism in the postslavery period. What is most striking is that he, a Frenchman, did so when American radicals and reformers almost unanimously neglected the issue. While the left of this period focused heavily on economic class issues and only gradually came to grasp the centrality of racism to the system of social injustice, liberals and "progressives" needed another half-century to discover the "American dilemma." Reclus, on the other hand, continued until his last days to develop the critique of racism that he had begun in the 1850s. He describes the American system of racial segregation with an acute sense of moral outrage. He recounts the harsh punishment, tortures, and murders committed against blacks who offended the mores of racist communities, noting that such "horrible practices" were so common that they had taken on the force of "local law."[68]

While Reclus deserves credit for such advanced views, certain serious limitations of his outlook must also be noted. Despite his fierce anti-racism and his appreciation of diverse cultures and peoples, he was not entirely successful in overcoming the Eurocentric ideology of his time. Especially in his early work, one detects undertones of condescension, even when he enthusiastically praises non-Western cultures. Also, surprisingly, in view of his hatred for racism and his experience of living in the South, he shows little awareness of the contributions of black culture to American culture in general. In his view, blacks had been so "deracinated" and so Americanized "by language, education, ways of thinking, and even patriotism and all its prejudices" that their "originality within the whole of the nation" became "minimal."[69] He should perhaps be given credit for avoiding the perennial leftist pitfall of uncritically idealizing oppressed groups, but unfortunately, in this case, he sometimes goes to the opposite extreme.

Despite such lapses, Reclus' abhorrence of racism and his quest for understanding and mutual recognition between all ethnic groups and cultures persisted throughout his life. His efforts to transcend the prejudices of his age became consistently more successful, and, especially in his mature writing, he exhibited unusual openness and perceptiveness in examining the values and achievements of every society. After his death,

Kropotkin could say of him with justice that "in speaking of the smallest tribe, he always found a few words to inspire his reader with the feeling that all men are equal, that there are no superior or inferior races."[70]

Just as through most of its history the theorists of the left neglected the issue of racism, they also exhibited a very limited awareness of the central place of patriarchy in the system of domination. On this topic, Reclus is also rather exceptional, for not only did he challenge the patriarchal system explicitly in his theoretical analyses, but—as is even more unusual in his epoch—he also attempted to put theory into practice in his personal life. In accord with his repugnance for all hierarchical relations, he opposed the concept of male dominance and advocated sexual egalitarianism. He believed that one precondition for equality between the sexes is the practice of "free unions" between men and women. In describing such unions, he states that "the normal, spontaneous family must be based solely on affection and on free affinity: Everything related to the family that arises out of the force of prejudice, the intervention of the law, or financial interests should disappear since it is essentially corrupting. Here, as in every other area, freedom and natural impulses are the basis of life."[71] In his view, such a union at its best is deeply fulfilling on many levels and contributes to the ongoing self-development of each partner. It "includes at once mutual passion, fervent friendship, perfect respect and the constancy of love that stems from continual transformation, from the renewal of each by the other until the end of their lives."[72] It is clear that Reclus' depiction of such a relationship was profoundly shaped by his own deeply fulfilling life with Fanny L'Herminez.

Compared to his ideal of "free unions, based only on mutual affection, self-respect and the dignity of others," he sees traditional marriage, authorized by the church and enforced by the state, as mere "matrimonial trafficking."[73] It is a morally debasing institution that lies at the core of the larger system of domination. Almost a half-century before Wilhelm Reich's revolutionary analysis of the connection between the authoritarian family and the authoritarian state, Reclus made strikingly similar claims. He argues that "it is certain that familial associations, whether manifested in polygyny, polyandry, monogamy, or free unions, exercise a direct influence on the form of the state through the effects of their ethics. What one sees on a large scale parallels what one sees on the small scale."[74] Though he, like other anarchist theorists, emphasizes the strong determining influence of the state on all other forms of domination, he is unusual in placing such a heavy emphasis on the correlatively powerful effects of family relationships on the state and other oppressive institutions. While not underestimating the evils of political coercion, he recognizes the ultimately greater force of psychological coercion operating in the context of the most intimate relationships. He notes the connection between the system of political authority and that prevailing in the family, and he remarks that the former is "ordinarily in lesser proportions, for the government is incapable of pressuring widely dispersed individuals in the way that one spouse can pressure the other who lives under the same roof."[75]

Reclus' views on marriage represent an important way in which he breaks fundamentally with the mainstream of modernist social thought, which tends to accept the division between the public and private spheres as autonomous realms. He believes that a free society can exist only if the principles of freedom that are to guide society are put into practice in the most intimate and personal details of life. His outlook anticipates the feminist interpretation of personal life as being eminently political and the post-1968 movement for the "liberation of everyday life." It also situates him in some ways closer to the tradition of utopian communitarianism than to the mainstream of modern anarchist and socialist political theory. The utopians have been among the few who have long taken questions of personal life seriously, while more conventional political radicals have usually seen changes in this realm as merely "superstructural" or have relegated them to the postrevolutionary era. Reclus saw an immediate and thoroughgoing change in personal relationships as a necessary precondition for liberatory social transformation. He comments that "it is above all within the family, in a man's daily relationships with those close to him, that one can best judge him. If he absolutely respects the liberty of his wife, if the rights and the dignity of his sons and daughters are as precious to him as his own, then he proves himself worthy of entering the assembly of free citizens. If not, he is still a slave, since he is a tyrant."[76]

In attempting to undermine the foundations of patriarchy and to demythologize it, Reclus looks back to the beginnings of human society. He was unusual for his time in his willingness to recognize the powerful contribution of women to the origins of civilization. According to his revisionist account of history, the institution of "maternity" (that is, of matricentric and matrilineal practices) arises "in the midst of primitive barbarism" and gives "the first impulse to the future civilization" by uniting the members of primitive bands around the maternal hearth and socializing them.[77] He believes that the role of women across the entire history of society has been vastly underestimated, both by scholars and in the popular mind. He notes that there is no lack of examples in history "of women who were veritable chiefs," that "diverse tribes have recognized absolutely the supremacy of women," and that "other tribes in which men have exercised power have adhered to the maternal family line."[78] Through such examples he seeks to destroy the myth of the universality and, by implication, the natural necessity, of patriarchy.

Reclus argues that the significance of women in the social institutions of many societies has been vastly underestimated. One of the most important areas that has been neglected in this way is economics. He notes that in societies where horticulture has been the prerogative of women, they have had "the useful role par excellence in the general economy of the tribe,"[79] and their labor has been the most secure source of food for the group. In such societies, "the general prosperity depends absolutely on capable management by the mothers, and on the spirit of order, peace and harmony that they introduce into the household."[80] Furthermore, in these cultures, the feminine influence is decisive for determining the values of the group, as "the natural affection that they bestow on the children gathered around them develops into a kind of religion."[81] Reclus also stresses the fact that, contrary to general misconceptions, women have often possessed powerful political authority in such communities. "No decision can be made without their being consulted beforehand. As the absolute dispensers of familial fortune, they come to be the regulators of all social and political affairs. Although the males are stronger, they bow before the moral sovereigns."[82] According to Reclus' analysis, even when in certain societies males performed the functions that from our perspective seem to be of greatest importance (e.g., the nominal "chief" may have been male), this does not necessarily indicate male social dominance. In such societies, even the exclusively male functions were subject to strong female influence, other functions of equal or greater importance were directly in the hands of women, and—what is most important—feminine and maternal values thoroughly pervaded the culture.

In his discussions of such societies, Reclus often refers to the "matriarchal family." This usage is a bit disconcerting since he purports to make an "anarchic" critique of all forms of social domination, yet we find him praising the superiority of another "archy." However, he recognizes that the concept of "matriarchy" would lead to confusion if taken in its literal etymological sense. He observes that in the kinship systems that are given this label the mother does not actually "rule." He notes that in fact the very significant maternal power that exists in such societies has sometimes been compatible even with "brutality" by the father, and with situations in which he is "the incontestable master" of the family.[83] Reclus is not describing a supposed system of female dominance. He does not attempt to invent a mythological "matriarchy" in which an imagined matriarchal power becomes the mirror image of historical patriarchal power. Rather, he seeks merely to show that patriarchy is not "inevitable," that women have often exercised authority in the most essential areas of social life, and that in doing so they have been the most powerful agents of "progress" and "civilization," in the best senses of those terms.

In addition to defending women's rightful place in history, Reclus vehemently supported their quest for social emancipation in his own day. In the strongest terms, he declares himself completely in accord with the feminist cause, asserting that "obviously, all of the claims of women against men are just: the demands of the female worker who is not paid at the same rate as the male worker for the same labor, the demands of the wife who is punished for 'crimes' that are mere 'peccadilloes' when committed by the husband, and the demands of the female citizen who is barred from all overt political action, who obeys laws that she has not helped to create, and who pays taxes to which she has not consented."[84] In short, women are oppressed not only in the domestic sphere but in the economic, social, and political ones also, and complete justice and equality must be achieved in all these areas.

But although Reclus is in sympathy with all the goals of feminism, he does not approve of all feminists. He is disturbed that some middle-class feminists seem concerned only with their own oppression and exhibit disdain for the working class. He laments the fact that they fail to see that "their cause merges with that of all oppressed people, whoever they may be."[85] Reclus' comments are echoed today by radical feminists who criticize liberal feminism for focusing on issues such as "the glass ceiling" that affects upwardly mobile, more privileged women while neglecting the suffering and oppression of working-class and poor women. Reclus celebrates "the heroism of brave women who go to the prostitutes to join them in solidarity to protest the abominable treatment to which they have been subjected, and the shocking bias of the law in favor of the corrupters and against their victims."[86] Reclus was far ahead of his time not only in speaking out for the cause of the most abused and abandoned of women but also in calling attention to the state's complicity with the men who exploit them.

Another of Reclus' views that has only recently begun to gain widespread sympathy is his firm belief that women are justified in striking back at their oppressors. He declares that as a result of the severe mistreatment to which they have been subjected, women have "an absolute right to recrimination, and the women who occasionally take revenge are not to be condemned, since the greatest wrongs are those committed by the privileged."[87] From Reclus' time to the present, few have defended retaliation by women except in the most extreme cases of abuse. However Reclus believed that in view of the brutality of the oppression of women and because the most oppressed found few advocates of their cause, overt rebellion was often an appropriate response. He deplores the fact that the cause of women is usually dominated by well-behaved, conventional personalities (moderate and liberal feminists, we would say today) who "naïvely petitions legislators and high officials, waiting for salvation through their deliberations and decrees," when in truth "freedom does not come begging, but rather must be conquered."[88] Reclus, like his great anarchist-feminist contemporary Emma Goldman, thought that women could only advance their cause effectively through direct action—in both the personal and social spheres.

Another area that was much neglected in Reclus' time, but about which he shows remarkable insight, is the question of the rights of children and the place of the young in society. His thoughts on this topic are closely related to his critique of patriarchy. In his view, just as it was necessary to break with the long tradition, running "from Aristotle, St. Paul and the Church Fathers" down to "the Fathers of the American Constitution," that saw slavery as a legitimate form of property ownership, so the equally ancient tradition that makes the father the proprietor of his children must be rejected.[89] Reclus proposes a new morality that will "recognize the free individual even in the newborn infant, and defend the child's rights in relation to all, including, first of all, the father."[90] A corollary of this view is that the repressive system of authoritarian education, which is an extension of the patriarchal family and the authoritarian state, must be abolished.

Reclus attacks the existing system of education as a process of training children to fit well into institutions based on egoism, domination, and unthinking obedience. Through its hierarchical structure this system teaches competition for personal advantage rather than cooperation in pursuit of the general good. The students, "from their first lesson, learn that they are rivals and combatants. They are told in every way that the prizes to be won are few in number, and that one must snatch them away from one's comrades, not only by superior talent, but, when possible, by trickery, by force, by cabals and schemes, by the basest sort of machinations, or by prayers to St. Anthony of Padua."[91] The goal is to convince the students that all sorts of future honors and benefits can be achieved if they are willing to fight for them and destroy others in the process. Humanity and solidarity are undermined for the sake of "these symbols."[92] Just as a system of libertarian education is necessary to create a community of free, compassionate, and cooperative human beings, a system of authoritarian education is essential to the production of a hierarchical society of dominant and submissive individuals.

Another area that for Reclus is fundamental to the creation of an authoritarian character structure is the system of repressive morality. One of the expressions of this morality that he finds most outrageous is the nudity taboo.[93] He believes that a free society can never be attained without the rehabilitation of the body and the complete affirmation of our physical being. From his perspective it is clothing, rather than the human body, that is the true scandal.

While Reclus has a long list of arguments against the evils of clothing, his principal objection is a moral one. He asserts that "it is from the point of view of moral health above all that the reintroduction of nude beauty is necessary."[94] A morality that "consists of repressing one's body" and pretending that one "no longer has any organs, results in constantly directing thought toward those things that should remain 'out of sight.' It is a phobia, madness, fierce lechery, the perversion of all the senses. It is lying, hypocrisy."[95] Opponents of nudity create a moralistic travesty of morality in which "normal acts become vicious" and "the source of life is corrupted," so that "from generation to generation the world is perverted."[96] From Reclus' point of view, the deleterious effects of clothing on individual character and social morality can hardly be overestimated.

He attacks what might be called the fetishism of clothing on grounds not only of morality but also of physical health. He claims that "without doubt, the skin regains its vitality and its natural activity when it is freely exposed to air, light, and the changing phenomena outdoors. Perspiration is not hampered; the functioning of the bodily organ is improved; it becomes at once firm and supple; it does not pale like an isolated plant deprived of sunlight."[97] While nudity, he contends, contributes to health, clothes are "nests of germs that cut us off from pure air and light, make us sickly and uncoordinated, turn our skin pale and cover it with ulcers, make lovers repulsive to one another, and sometimes make women sterile or doom them to give birth to weak and stunted infants!"[98] Reclus' feelings about clothing place him in sympathy with the dress reform movement of his time, which sought to liberate both women and men from restrictive and unhealthy clothing. For example, this movement blamed tight corsets, popular during the Victorian era, for a variety of health problems, including constriction and displacement of vital organs, atrophy of muscles, and risks to pregnant women and their fetuses.[99]

Reclus also objects to the obsession with clothing on the grounds of aesthetic appreciation. He cites travelers' accounts of the Polynesians as "the most beautiful of people" in their nudity before "the missionaries went on their rampage," and he notes the universal admiration for Greek artistic representations of the nude.[100] He believes that without the liberation of the body and the freedom to appreciate the body, the full development of art is not possible. He also believes that clothing is destructive of beauty in everyday life. As a result of clothing and the dictates of fashion, "natural curves are replaced by rows of buttons, and by skirts and blouses."[101]

While this spirited tirade certainly makes some excellent points about repression and hypocrisy, it also exhibits a paradoxical relationship between his nudism and the very Puritanical spirit that he attacks. He praises nudity, quite consistently, in the name of free expression and healthy self-affirmation. However, the seemingly obvious possibility that clothing and self-decoration might also be positive forms of individual and social self-expression is rather dogmatically overlooked. Clothing for Reclus can only be seen as an expression of corrupt society, with its class hierarchy and sexual repression. "The artifice of dress and finery is one that leads most—through foolish vanity, the slavish spirit of imitation, and above all, the thousand ingenuities of vice—to the general corruption of society. . . . Nude beauty purifies and ennobles; clothing, insidious and deceptive, degrades and perverts."[102]

At this point, the radicalism of Reclus' position is undermined by the reactive, absolutist quality of his response to a corrupt and hypocritical society. Moreover, his own viewpoint borders on a form of naturist Puritanism when, in criticizing the distorted eroticism of a repressive society, he comes close to purging both nudity and clothing of any positive erotic potential. He certainly deserves recognition for his courageous application of the idea of human liberation to the body, a subject neglected by most radical theorists (other than a few "utopians") until recent times. However, his idealization of nudity sometimes falls into the kind of naïve, reactive naturalism that has plagued ecological thought even up to the present.

An institution that Reclus sees as closely allied to authoritarian morality is authoritarian religion. As was discussed in considerable detail earlier, he sees religion as having both progressive and regressive aspects. He holds that the founders of the great religions had metaphysical and moral insights that conflicted starkly with the later religious institutions that he attacks so vehemently. These original insights often had subversive and even revolutionary implications that had to be negated in order for religion to be transformed into an ideology at the service of patriarchy, state power, and economic exploitation. It is thus by betraying its own egalitarian and libertarian premises that religion becomes one of the most powerful cultural and psychological supports for oppression and domination. Reclus believes that institutionalized religion has carried out this authoritarian role very successfully across history. As he sums up the tragic and brutal history of institutionalized religion in one of his letters, "the fear of God is the beginning of all servitude and all depravity."[103]

As has been mentioned, Reclus believes that the inexorable progress of science will progressively destroy the ideological basis of religion and that it will therefore be in a state of constant retreat before the imperious forces of modernity. In fact, he contends that although religion was once a powerful form of social domination in its own right, by his own time it had already lost much of its hold on the masses and was becoming primarily a system of social convention. This view, which exaggerates the decline of religious institutions, seems to be shaped strongly by Reclus' experience of the Catholic cultures of southern Europe. He observes that in these cultures "the interests of property, capital, parasitism, and everything of this sort demand the prescribed practice of the Catholic religion, and millions of people conform to this obligation, carried out without the least sincerity."[104] He contends that not only in southern Europe but in much of the world, religion had been reduced to such superficial practice and no longer consisted of deeply held beliefs. He judges, however, that even as a form of social convention religion would continue for some time to function as an important support for other forms of domination, such as the state and capitalism, that are now in a period of historical ascendancy.

It is likely that Reclus would have no difficulty recognizing the authentically progressive nature of liberation theology and other religiously based social justice movements today. Also, the fact that religion continues to reinforce various systems of political, economic, and cultural power would be fully in accord with his expectations. On the other hand, in view of his thesis that the advances of science and general enlightenment would result in the progressive decline of religious institutions, he would no doubt be surprised at the enduring strength of highly repressive, doctrinaire religious movements, at the global resurgence of fundamentalism, and at the ability of religion to function at times as a relatively autonomous form within the larger system of domination.

Despite certain limitations that have been mentioned, Reclus' critique of domination is an analysis with a consistency and comprehensiveness that are impressive. Not only is it broad enough to encompass many of the major institutions of society, including politics, economics, technology, sex roles, family structure, education, morality, and religion, but it is also deep enough to extend from the level of vast social institutions to the most intimate areas of personal life and human relationships. For this, Reclus must be recognized as a groundbreaking social theorist whose thought is a landmark in the development of the critical theory of society.

7

The Legacy of Reclus: Liberty, Equality, Geography

We will conclude by returning to the beginning—to the beginning of *Man and the Earth*, and to the beginning of Reclus' entire problematic, both as a thinker and as a human being. He begins his great work of social geography with the image of human hands holding the earth, an image that reveals much about what is most essential to his outlook and about the nature of the ecological imaginary implicit in his work. First, the earth is held aloft almost as if it were a sacramental object. It is presented as the object of awe, reverence, deep love, and respect. Second, and perhaps more obviously at first glance, the image depicts the earth as being "in the hands" of a personified humanity.[1] The image thus points to our responsibility for the destiny of the earth, and to our need to achieve the collective self-consciousness symbolized by this image of humanity.

These two aspects of the image capture very well the two imaginary poles of Reclus' thought: the ecological imaginary that is embodied in his social geography, and the anarchistic imaginary that is embodied in his politics. First, we are challenged to develop a deeper respect, reverence, and love for nature, for the earth, and for all the beings that share the planet, including humanity in all its diversity. Secondly, we are challenged to express these feelings in fully engaged, transformative activity, as we carry out our responsibilities toward all that we find in our hands.

We have seen that Reclus is distinctive among modern social theorists for the central place of love in his conception of social transformation. This is a powerful dimension of his thought that speaks directly to the crisis of our age. As valid as the message of social and environmental justice may be, it is a message that has no meaning for those who do not care deeply about humanity and the earth. For this reason the revolutionary social transformation that Reclus calls for implies not only an objective revolutionizing of social institutions but also a revolution in subjectivity.

In such a revolution of the spirit, human beings rediscover and experience more vividly their connectedness with others and with nature.

While there is perhaps something implicit in our "egocentric predicament" that will always incline us toward egoism, it is the patriarchal, authoritarian, power-based institutions of society that transform this inclination into an egoistic rage against the other human being and against nature itself.[2] Reclus' importance lies not least of all in the fact that he combined a critically holistic, communitarian vision of society and nature with insight into the social barriers that prevent human beings from relating themselves to these greater wholes in a practical, socially transformative manner. While his dialectical, holistic view of humanity-in-nature helps diagnose our egoistic, autistic malady, his analysis of the institutions and ideologies of domination—capitalism and class domination; nationalism, statism and technobureaucracy; patriarchy and sexism; racism and ethnic oppression; speciesism and the domination of nature—helps show what must be changed if our autism is to be cured. The way out of our egocentric impasse is a process of self-transformation that is identical with a process of social evolution/revolution.

Reclus' most enduring legacy is his contribution to our growing self-knowledge as human beings and as planetary beings, and to the reemergence of the spirit of hope and of creative action. His significance lies in his place on the way to the great convergence of reason, passion, and imagination—*logos*, *eros*, and *poesis*. It comes from his work in preparing for the day when poetry, myth, and narrative enter fully into a dialectic with reason and experience. Reclus spoke fervently of revolution, which in his liberatory imaginary was the most inspiring image of hope. But he also contributed powerfully to a vision of the future rooted in the much more ecological image of "regeneration." He pointed toward the regeneration of a rich, highly individualized yet social self; the regeneration of a free, cooperative community; and the regeneration of a holistically diversified, dynamically balanced, creatively evolving earth. This is the utopian and topian vision in which Reclus' version of the human story and the earth story culminate—a realm of freedom encompassing humanity and the whole earth; an end to the domination of humanity and of all other beings on the planet; and humanity's final realization of its harmonious, integral place in nature.

Liberty, Equality, Geography!

PART II

Selected Writings of Elisée Reclus

8

The Feeling for Nature in Modern Society (1866)

The following discussion is taken from the relatively early article "Du sentiment de la nature dans les sociétés modernes," which was published in *La Revue des Deux Mondes* 63 (May–June 1866): 352–81. It is noteworthy as an example of Reclus' view of nature in his earlier work. Most of the social analysis in the three-part essay appears in the third section (371–81), which is translated here.

It becomes ever more essential to expand and refine our feeling for nature as the multitude of men who are exiled from the countryside by force of circumstances increases daily. Pessimists have long feared the ceaseless growth of large cities. Still, they seldom realize how rapidly future populations will be able to move toward preferred centers.

It is true that the colossal Babylons of the past also gathered within their walls hundreds of thousands or even millions of inhabitants. The natural interests of commerce, the despotic centralization of all power, the scrambling for favors, and the pursuit of pleasure made these powerful cities as populous as entire provinces. But factors such as slow transportation, the flooding of a river, bad weather, the delay of a caravan, a raid by an enemy army, or a tribal uprising could result in provisions sometimes being delayed or halted. The great city, in the midst of all its splendors, found itself in constant danger of starvation. Moreover, during periods of relentless war, these enormous capitals always ended up as an arena for an immense slaughter, and sometimes the destruction was so complete that the ruin of a city meant the end of a people. Even quite recently, we were able to see, through the example of several cities in China, what fate could befall great urban centers under the sway of ancient civilizations. The powerful city of Nanjing was reduced to a heap of ruins, while Wuchang, which about fifteen years ago appeared to be the most populous city in the entire world, lost more than three-fourths of its inhabitants.

While traditional causes of population shifts to large cities still operate, there are now other no less powerful causes that relate to the whole of modern progress. Transportation routes, canals, secondary roads, and railroads radiate in increasing numbers from important centers and surround them with an increasingly dense network of links. Today's transportation is so smooth that during a single day the railways can deliver five hundred thousand persons to the streets of London or Paris, and in anticipation of a simple holiday, a wedding, a funeral, or the visit of a celebrity, millions have sometimes swelled the fluctuating population of a capital. And provisions can be transported just as easily as travelers. From the surrounding countryside, outlying parts of the country, and all corners of the world, commodities flow by land and sea toward these enormous stomachs that endlessly consume more and more. If it were necessary for the demands of its appetite, London could have more than half of the earth's produce transported to it in less than a year.

This is certainly an enormous advantage that the large cities of antiquity did not have, yet the revolution in social practices brought about by railroads and other modes of transportation has hardly begun. After all, what is an average of two or three trips per year for each inhabitant of France, especially when a brief excursion of fifteen minutes to the suburbs of Paris or some other large city is considered a trip for the statistics? Each year, the multitude of travelers will doubtless increase in enormous proportions, and all expectations will probably be surpassed, as they have been since the beginning of the century. Thus the amount of travel in London alone is currently as great in a single week as it was during an entire year in all of Great Britain around 1830. Thanks to the railroads, regions are constantly becoming smaller. One can even mathematically calculate the rate at which this shrinking of the land is taking place merely by comparing the speed of locomotives to that of the stagecoaches and rickety carriages that they replaced. For his part, man turns his back on his native soil more and more easily. He becomes a nomad—not like the shepherds of the past, who always followed their usual paths and never failed to return periodically to the same pastures with their flocks, but in a manner much more complete since he indiscriminately heads in one direction or another, wherever his interest or desire impels him. A very small number of these voluntary exiles return to die in their native land. This endlessly growing migration of peoples is now taking place by millions upon millions, and it is precisely toward the most populous human anthills that the great multitude of immigrants makes its way. From an ethnological point of view, the fearsome invasion of Frankish warriors into Roman Gaul was perhaps not as important as the silent migration of street sweepers from Luxembourg and the Palatinate who each year swell the population of Paris.

To get an idea of what the great commercial cities of the world could become if the causes of growth are not sooner or later counterbalanced by opposing factors, one can simply observe the enormous importance of cities in modern colonial societies relative to villages and isolated households. The populations of these regions, released from the bonds of custom, and free to congregate as they please, with no motive except their own will, amass overwhelmingly in the cities. Even in specifically agricultural settlements such as the young American states of the Far West, the regions of La Plata, Queensland in Australia, and the North Island of New Zealand, the urban population surpasses that of the countryside. On average it is at least three times greater and constantly increases in proportion to the development of commerce and industry. In settlements such as Victoria and California, where specific factors such as gold mines and great commercial advantages attract multitudes of speculators, the concentration of city dwellers is greater still. If Paris were to France what San Francisco is to California and what Melbourne is to sunny Australia, the "big city" would really live up to its name, having no less than nine to ten million people. Clearly, it is in all these new countries, where civilized man has only recently established himself, that one can see the external expression of the ideal of nineteenth-century society: no obstacle prevented the newcomers from spreading out in small groups over the entire region, yet they preferred to gather in vast cities. The contrast between Hungary or Russia and any modern colony such as California illustrates how great a gap of centuries separates countries whose populations are still distributed as in the Middle Ages from those where the phenomena of social affinity developed by modern civilization can have free play. On the plains of Russia and in the Hungarian *puszta*, there are hardly any true cities, but only more or less large villages. The capital cities are administrative centers, artificial creations that the inhabitants could easily do without and that would immediately lose a sizable share of their importance if the government did not maintain a factitious life there at the expense of the rest of the nation. In these countries the working population is composed of farmers, and the cities exist only for office workers and men of leisure. By contrast, in Australia and California the countryside is never more than a suburb, and its inhabitants, shepherds and farmers, have their minds on the city. They are speculators who have temporarily withdrawn from the great commercial center for the sake of their business but who will inevitably return to it. Doubtless, the Russian peasants who are now so firmly rooted in their native soil will sooner or later discover how to free themselves from the fields on which only yesterday they were subjugated. Like the British and the Australians, they will become nomads and make their way to the big cities, beckoned by commerce and industry and compelled by their own ambition to see, to know, or to improve their condition.

The complaints of those who lament the depopulation of the countryside cannot stop the movement. Nothing will stop it, and all the outcry is useless. Thanks to easier and cheaper travel, the tenant farmer has gained the fundamental liberty to "come and go," from which all other liberties eventually proceed, and he follows his natural inclination when he heads for the crowded city, about which he has heard so many wonderful tales. Sad and joyful at the same time, he bids farewell to the lowly hovel of his birth to gaze upon the miracles of industry and architecture. Although he gives up the regular and dependable wages from his manual labor, perhaps he will succeed, like so many other sons of his village, in becoming comfortably well-off or even wealthy. And if he returns home one day it will be to build a castle in place of the squalid dwelling where he was born. However, very few immigrants realize such dreams of fortune, though many find poverty, disease, and a premature death in the big cities. But at least those who survive are able to broaden the horizon of their ideas. They have seen regions that differ from one another, developed themselves through contact with other men, and become more intelligent and educated, and all these individual advancements constitute an invaluable asset for society as a whole.

In France, we know how rapidly the phenomenon of the migration of rural populations toward Paris, Lyons, Toulouse, and the large seaports takes place. All population growth occurs in these centers of attraction, whereas the number of inhabitants in most of the small towns and villages remains stationary or even declines. More than half of the *départements* are becoming less and less populated, and one can be cited, that of Basses-Alpes, which since the Middle Ages has undoubtedly lost a good third of its inhabitants. If one also takes into account visits and temporary migrations, which necessarily produce an increase in the fluctuating population of the big cities, the results are even more striking. In the Pyrenees of Ariège, there are certain villages that all the inhabitants, both men and women, abandon in the winter in order to go to the cities of the plains. Finally, most Frenchmen who are in business or who live off their investments—not counting the multitudes of peasants and workers—are certain to visit Paris and the main cities of France. And it has been a very long time since, in remote provinces, a wayfaring laborer was named after the large city in which he lived. The same social phenomena are occurring in England and Germany. Although in these two countries the excess of births over deaths is much greater than in France, some agricultural areas such as the duchy of Hesse-Cassel and the county of Cambridge are also losing population to the large cities. Even in North America, where the population is increasing at an astonishing rate, a great number of agricultural areas in New England have lost a large proportion of their inhabitants because of a double migration: on the one hand, there is a movement toward the regions of the Far West, and on the other, toward the coastal commercial cities of Portland, Boston, and New York.

However, it is a well-known fact that in the cities the air is full of deadly substances. Although the official statistics on this matter are not always as candid as we would like them to be, it is nonetheless certain that in all countries of Europe and America, the average life-span among rural populations exceeds that of the city dwellers by several years. Immigrants who leave their native soil for the narrow and foul-smelling streets of a big city could calculate in advance the approximate extent to which they are shortening their lives according to the laws of probability. Not only does the newcomer suffer personally and risk an early death, he also dooms his descendants. It is known that in large cities such as London and Paris the life force is quickly exhausted, and that no bourgeois family living there survives beyond the third or at most the fourth generation. If the individual can resist the deadly effects of his environment, his family will still succumb in the end, and without the continuous migration of country people and foreigners who march happily to their death, the capital cities could not recruit their enormous populations. The city dweller's character becomes refined, but the body weakens and the springs of life dry up. Likewise, from an intellectual point of view, all the brilliant faculties developed by social life are at first overstimulated, but the mind gradually loses its powers. It becomes weary and finally declines prematurely. The street urchin of Paris, compared to the young peasant, is certainly a being full of life and high spirits. But is he not the brother of the pale hoodlum who can be compared physically and morally to sickly plants vegetating in dark cellars? In fact, it is in the cities, especially those most renowned for their opulence and civilization, that one finds the most degraded of all men. They are poor beings without hope, whom filth, hunger, coarse ignorance, and general contempt have placed far below the happy savage wandering through forests and mountains. One finds the rankest abjection side by side with the most magnificent splendor. Not far from museums where the beauty of the human body is displayed in all its glory, spindly children warm themselves in the foul atmosphere emanating from sewers.

If steam power brings endlessly growing crowds to the cities, it also brings back to the countryside an ever-growing number of city dwellers who go to breathe the open air for a while and refresh their minds among flowers and greenery. The wealthy, free to create leisure time as they please, can escape their occupations and the weary pleasures of the city for months at a time. There are even those who live in the countryside and make only fleeting appearances at their city residences. As for the workers of all types, who cannot leave for long periods because of the demands of everyday life, most manage nevertheless to take enough time off from their jobs to visit the countryside. The most fortunate among them take weeks of vacation, which they spend far from the capital, in the mountains or at the seashore. Those who are the most enslaved by their work content themselves with an occasional escape from the narrow horizons of their accustomed streets for a few hours. Naturally, they happily take advantage of their holidays when the weather is mild and the sky is clear. At such times, every tree in the woods near the big cities shelters a happy family. A considerable proportion of merchants and clerks, especially in England and America, bravely establish their wives and children in the countryside and sentence themselves to traveling twice per day the distance that separates the sales counter from the domestic hearth. Thanks to the speed of transportation, millions of men can lead the double lives of city and country dweller, and each year, the number of persons who thus divide their lives constantly grows. Each morning, hundreds of thousands converge on London to plunge into the whirlwind of business in the big city, and then return each evening to their peaceful homes in the verdant suburbs. The city, the true center of the business world, is losing its residents. By day, it is the most active human beehive; by night, it is a desert.

Unfortunately, this reflux from the cities toward the outskirts does not occur without defacing the countryside. Not only does debris of all sorts clutter the intermediate space between city and field, but even worse, speculators grab up all the charming sites in the vicinity, divide them into rectangular plots, enclose them with monotonous walls, and then build hundreds and thousands of pretentious little houses. To pedestrians wandering along the muddy roads in this would-be countryside, the only nature in evidence is the trimmed shrubs and clumps of flowers glimpsed through the fences. At the seashore, many of the most picturesque cliffs and charming beaches are snatched up either by covetous landlords or by speculators who appreciate the beauties of nature in the spirit of a money changer appraising a gold ingot. In frequently visited mountainous areas, the same mania of appropriation seizes the inhabitants. Landscapes are carved up into squares and sold to the highest bidder. Each natural curiosity, be it rock, grotto, waterfall, or the fissure of a glacier—everything, even the sound of an echo—can become individual property. The entrepreneurs lease waterfalls and enclose them with wooden fences to prevent non-paying travelers from gazing at the turbulent waters. Then, through a deluge of advertising, the light that plays about the scattering droplets and the puffs of wind unfurling curtains of mist are transformed into the resounding jingle of silver.

Since nature is so often desecrated by speculators precisely because of its beauty, it is not surprising that farmers and industrialists, in their own exploitative endeavors, fail to consider whether they contribute to defacing the land. Certainly the "sturdy plowman" cares very little for the charm of the countryside and the harmony of the landscape, so long as the soil produces abundant harvests. Walking around the thickets at random with his ax, he cuts down trees that are in his way and shamefully mutilates others, giving them the appearance of posts or brooms. Vast regions which formerly were beautiful to behold and enjoyable to travel through are completely spoiled, and one actually experiences disgust upon seeing them. Moreover, it often happens that the farmer, as lacking in science as he is in love of nature, errs in his calculations and causes his own ruin through certain changes that he unwittingly introduces into the environment. Similarly, it matters little to the industrialist, operating his mine or factory in the middle of the countryside, whether he blackens the atmosphere with fumes from the coal or contaminates it with foul-smelling vapors. In Western Europe, not to mention England, there are a great many industrial valleys whose thick air is almost unbreathable to outsiders. The houses there are filled with smoke, and even the leaves on the trees are coated with soot. The sun almost always shows its yellowish face through a thick haze. As for the engineer, his bridges and viaducts always look the same, whether on the flattest of plains or in the gorges of the steepest mountains. He is concerned not with making his work harmonious with the landscape, but solely with balancing the thrust and resistance of his materials.

Certainly, man must take possession of the earth's surface and know how to utilize its forces. However, one cannot help lamenting the brutality with which this process is carried out. And so when the geologist Marcou[1] informs us that Niagara Falls has noticeably decreased in flow and lost its beauty since it was diverted to operate factories on its banks, we think sadly of a time not long ago when the "thunderous waters," unknown to civilized man, tumbled freely over the high cliffs between two walls of rock completely covered with large trees. Similarly, one wonders whether the vast prairies and wild forests, where one can still imagine seeing the noble figures of Chingachgook and Leatherstocking,[2] could have been succeeded by something other than fields of equal size, all aligned with the points of the compass, in accordance with the land survey, and enclosed uniformly with fences of a standard height. Wild nature is so beautiful. Is it really necessary for man, in seizing it, to proceed with mathematical precision in exploiting each new conquered domain and then mark his possession with vulgar constructions and perfectly straight boundaries? If this continues to occur, the harmonious contrasts that are one of the beauties of the earth will soon give way to a depressing uniformity. Since society is increasing its population by at least ten million per year and has at its disposal through science and industry forces that are growing at a phenomenal rate, it is marching rapidly toward the conquest of the entire surface of the planet. The day is approaching when there will remain no region on any continent that has not been visited by a civilized pioneer, and sooner or later, the effects of human labor will extend to every point on the surface of the earth. Fortunately, a complete alliance of the beautiful and the useful is possible. It is precisely in the countries where industrialized agricultural is most advanced—in England, Lombardy, and certain parts of Switzerland—that those who exploit the soil know how to make it produce the highest yields while at the same time respecting the charm of the landscape, or even adding artfully to its beauty. The marshes and bogs of Flanders, transformed by drainage into extremely fertile countryside; the rocky Crau, changed into a magnificent prairie thanks to irrigation canals; the rocky slopes of the maritime Apennines and Alps, covered from base to summit with the foliage of olive trees; and the reddish peat bogs of Ireland, replaced by forests of larch, cedar, and silver fir—are these not admirable examples of this power by which the farmer exploits the land for his benefit while at the same time rendering it more beautiful?

The question of knowing which of the works of man serves to beautify and which contributes to the degradation of external nature can seem pointless to so-called practical minds; nevertheless, it is a matter of the greatest importance. Humanity's development is most intimately connected with the nature that surrounds it. A secret harmony exists between the earth and the peoples whom it nourishes, and when reckless societies allow themselves to meddle with that which creates the beauty of their domain, they always end up regretting it. In places where the land has been defaced, where all poetry has disappeared from the countryside, the imagination is extinguished, the mind becomes impoverished, and routine and servility seize the soul, inclining it toward torpor and death. Throughout the history of humanity, foremost among the causes that have vanquished so many successive civilizations is the brutal violence with which most nations have treated the nourishing earth. They cut down forests, caused springs to dry up and rivers to overflow, damaged environments, and encircled cities with foul-smelling marshes. Then, when nature thus desecrated turned hostile toward them, they came to hate it, and, unlike the savage, who could immerse himself in the life of the forest, they increasingly allowed themselves to succumb to the stupefying despotism of priests and kings. "The great estates have ruined Italy," said

Pliny, and it must be added that these great estates, cultivated by slaves' hands, defaced the land like leprosy. Historians, struck by the astonishing decline of Spain since Charles the Fifth, have tried to explain it in various ways. According to some, the principal cause of that nation's downfall was the discovery of gold in America; others claim that it was the religious terror organized by the "holy brotherhood" of the Inquisition, the expulsion of the Jews and the Moors, and the bloody autos-da-fé of the heretics. They have also blamed the fall of Spain on the unfair tax of the alcabala and the despotic centralization in the French manner.[3] But did the Spanish passion for cutting down trees due to their fear of birds, "por miedo de los pajaritos," contribute nothing to this terrible decline? The earth, yellow, rocky, and naked, has taken on a repugnant and fearsome appearance: the soil is impoverished, and the population, which has been decreasing for two centuries, has to an extent lapsed into barbarism. The little birds are avenged.

Therefore, we must now enthusiastically welcome the generous passion that induces so many men (and we declare them to be the best among men) to traverse virgin forests, beaches, and mountain gorges, in short, to visit nature in all regions of the earth that have retained their original beauty. Threatened with intellectual and moral decline, one feels the need to see the great sights of the earth in order to counterbalance at all costs the vulgarity of all the ugliness and mediocrity that narrow minds view as evidence of modern civilization. The direct study of nature and the contemplation of its phenomena must become for all well-rounded men one of the fundamental elements of education. It is also essential for each individual to develop muscular dexterity and strength so that he can enjoy climbing to the peaks of mountains, look fearlessly into abysses, and keep in his entire physical being that natural balance of forces without which one can perceive the most beautiful settings only through a veil of sadness and melancholy. Modern man must unite in his being all of the virtues of those who have preceded him on earth. Without giving up any of the great privileges that civilization has conferred on him, neither must he lose any of his ancient strength, nor allow himself to be surpassed by any savage in vigor, dexterity, or in knowledge of natural phenomena. In the splendid epoch of the Greek republics, the Hellenes undertook nothing less than to make their children heroes through grace, strength, and courage. In the same way, it is by awakening in the younger generations all of the qualities of manliness and by bringing them back to nature and making them come to grips with it that modern societies can be insured against all decline through the regeneration of the race itself.

Rumford said a long time ago that "one always finds in nature more than one is looking for."[4] Whether the scholar examines clouds or stones, plants or insects, or whether he goes further and studies the general laws of the world, he continually discovers unexpected wonders everywhere. The artist who seeks out beautiful landscapes encounters a continual feast for the eyes and mind. The industrialist who tries to make use of what the earth produces inevitably sees around him unutilized riches. As for the simple man who is content to love nature for itself, he finds in it his joy, and when he is unhappy, his sorrows are at least mitigated by the sight of the wild countryside. Certainly, outcasts or even those poor déclassés who live like exiles in their own homeland always feel isolated, unknown, and friendless, even in the most charming settings, and they suffer the constant ache of despair. However, in the end they also experience the gentle influence of their environment, and their most intense bitterness gradually changes into a sort of melancholy that allows them to comprehend, with a sensibility refined by sadness, all that the earth has to offer in grace and beauty. Even more than those who are happy, they know how to appreciate the rustling of leaves, the songs of birds, and the murmur of springs. And if nature has the power to console or to strengthen individuals, what could it do over the course of centuries for whole peoples? Without a doubt, magnificent vistas greatly contribute to the qualities of mountain populations, and it is no mere figure of speech to call the Alps the boulevard of liberty.

9

To My Brother the Peasant (1893)

In 1873, Reclus wrote an article entitled "Quelques mots sur la propriété" for *L'Almanach du peuple*. He later revised and expanded it, publishing it as a pamphlet under the title *A mon frère le paysan*. In his "Biographie d'Elisée Reclus" in *Les Frères Elie et Elisée Reclus* (Paris: Les Amis d'Elisée Reclus, 1964), Paul Reclus writes that "it was translated into a dozen European languages, even including two dialects of Breton" (91). While this small work is a classic of anarchist propaganda and possesses all the rhetorical qualities appropriate to the genre, it is also of interest for its comments on the relationship between capitalism and technological rationality.

"Is it true," you ask me, "that your comrades, the urban workers, are thinking of seizing the land from me, this sweet land that I love and that bears me crops? It does so very meagerly, I'll admit, but nonetheless it bears them. It has fed my father and my father's father. And won't it provide a little bread for my children, too? Is it true that you want to seize this land from me?"

"No, brother, it's not true. Because you love the soil and cultivate it, the harvest indeed belongs to you. You are the one who produces the bread, and no one has any more right to it than you, the wife who shares your lot, and the child born of your union. Keep your fields in peace, keep your spade and plow to turn the hard soil, and the seed with which to make it fruitful. Nothing is more sacred than your labor, and a thousand curses on whoever would seize the land that has become nourishing through your efforts!"

But what I say to you, I do not say to others who claim to be farmers but who in reality are not. Who are these self-styled workers, these fertilizers of the soil? One of them is born a great lord. As he is being placed in his cradle, wrapped in fine wools and soft, beautiful silks, the priest, the judge, the lawyer, and other dignitaries arrive to greet the newborn as a future master of the earth. Sycophants, both men and women, hasten from all around to bring him presents of silver brocades and golden rattles. While he is being showered with gifts, pencil pushers are recording in great books that the little one is the owner of springs and rivers in one place, woods, fields, and prairies in another, and additional parks, fields, woods, and pastures in yet another. He owns property in the mountains and also in the plains. He is even the master of great underground domains in which men work by the hundreds and thousands. Some day, when he grows up, he may decide to visit what he inherited as his birthright, or perhaps he will never even bother to go and see all these things; however, he won't forget to have the produce gathered and sold. From all over, by highways and railroads, by riverboats and by oceangoing ships, he is brought large bags of money, the income from all his landholdings. So, when we have gained strength, will we leave all these products of human toil in the safe of the heir? Will we have respect for this property? No, my friends, we will take all of it. We will tear up the documents and maps, break down the doors of the chateaus, and seize the estates. "Work, young man!" we will tell him. "Work if you want to eat. None of this wealth belongs to you any more."

And what about the other lord, the one who was born poor, without a pedigree, and whom no flatterer came to admire in the rude cottage or garret of his birth, but who was lucky enough to become wealthy through his work, honest or otherwise? He did not have a clod of earth to call his own, but owing to speculation, savings, favors from the authorities, or good fortune, he knew how to acquire immense stretches of land, which he now encloses with fences and walls. He harvests where he has not sown the seed, and he gathers and eats the bread that another has earned through his labor. Will we respect this second kind of ownership, that of the nouveau riche who never works his land himself but who has it worked by slave labor and calls it his own? No, we will not respect this second kind of ownership any more than the first. Here again, when we have gained strength, we will seize these estates as well and tell the one who considers himself the master: "Stand back, upstart! Since you once knew how to work, get back to work! You will have the bread that you gain through your labor, but the land that others cultivate is no longer yours. You are no longer the master of bread!"

And so, yes—we will seize the land, but we will seize it from those who hold onto it without working it, in order to return it to those who do. However, the latter will not then be allowed in turn to exploit other unfortunate people. The amount of land to which the individual, the family, or the community of friends has a natural right is the amount that can be worked through individual or collective labor. As soon as a parcel of land exceeds the amount that they are able to cultivate, they would be wrong to claim this additional portion. Its use belongs to another worker. The boundary will be drawn in different ways between the various lands cultivated by individuals or groups, depending on the requirements of production. The land that you cultivate, brother, is yours, and we will help you to keep it by every means in our power. But the land that you do not cultivate is for a companion. Make room for him, for he, too, knows how to make the land fruitful.

But even though each of you has the right to your share of land, do you want to remain isolated? Completely alone, the small farmer, whether landowner or tenant, is too weak to struggle against a stingy nature and an evil oppressor at the same time. If he survives, it is through a feat of willpower. He must adjust to all the whims of the weather and submit a thousand times to voluntary torment. Whether in freezing cold or blistering heat, whether in rain or wind, there is always work to be done. If floods drown his harvest or heat scorches it, he sadly gathers whatever remains, which will not be enough to provide for him. When sowing time arrives, he withholds grain that could have been eaten and scatters it in the field. In his despair, he is left with a grim faith: if necessary, he sacrifices a portion of his meager harvest, trusting that after a harsh winter, treacherous spring, and burning summer, the grain will still come up again, doubling or tripling the seed, or perhaps even increasing it tenfold. What intense love he feels for this land, which brings him so much misery and toil, so much suffering from fears and disappointments, yet so much joy when he sees the undulating stalks full of grain! No love is stronger than that of the peasant for the soil he plows and sows, from which he is born and to which he will return! Yet so many enemies surround him and covet the land that he loves! The tax collector taxes his plow and takes some of his grain from him, the merchant seizes another portion, and finally the railroad defrauds him when transporting his produce. He is tricked from all sides. We have called out to him, "Don't pay the tax, don't pay the rent!" Nevertheless, he still pays, because he is alone, does not trust his neighbors, the other small farmers, whether landowners or tenants, and does not dare to consult with them. They are kept submissive by fear and disunity.

The peasants who have joined in a *Zadrougas* ("group of friends") or in a *mir* (little "universe") such as those in Russia or other Slavic countries are stronger against the common enemy—the state and the feudal lord. Their collective property is not divided into countless enclosures by hedges, walls, and ditches. They do not have to quarrel over the ownership of an ear of corn growing to the right or the left of the furrow. There are no bailiffs, attorneys, or notaries to regulate business between comrades.

After the harvest and before the time comes to begin their work again, they gather to discuss their common interests. The young man who has just married, and the family that has added a child or taken in a guest, explain their new situation and take a larger portion of the common resources in order to satisfy their increased needs. Boundaries are decreased or increased according to the availability of land and the number of members. Each cares for his field, happy to be at peace with his brothers, who work their share of the land, which has been apportioned to meet the needs of all. During emergencies, the comrades help each other out. If a fire devours one of the cottages, all participate in rebuilding it. If a gully erodes part of a field, another portion is granted to the holder of the damaged land. One person grazes the community's herds, and in the evening, the sheep and cows follow the road back to their stables without being driven. The commune is at once the property of each and of all.

But the commune, like the individual, is still very weak if it remains isolated. Perhaps it does not have enough land for all of its members, and all will have to suffer from hunger! It almost always finds itself in conflict with a lord richer than itself, who claims ownership of a certain field, forest, or pasture. It puts up a good fight, and if the lord were alone, the commune would indeed quickly triumph over this greedy and arrogant personage. But he is not alone—he has on his side the governor of the province, the chief of police, the priests, the judges, and the entire government with its laws and its army. Should he need them, he has at his disposal cannons to bombard anyone who would fight him for the contested land. And even though the commune might be absolutely in the right, it is certain that the powerful will prove it wrong. And as much as we have cried out to the commune, as we did to the individual tax victim, "Don't give up!" it must also succumb, a victim of its isolation and weakness.

Thus all you small landowners, whether isolated or joined in communes, are indeed weak against those who try to enslave you—the land grabbers who are after your small plot of land and the authorities who try to take all the income from it. If you do not know how to join together, you will soon share the fate of millions upon millions of men who are already stripped of all rights to sow and reap and who live as wage slaves. They find work when the bosses are interested in giving it to them, and are always obliged to beg in a thousand ways, sometimes asking humbly to be hired, sometimes even holding out their hands to plead for a meager pittance. They have been deprived of land, and you might be among them tomorrow. Is there really such a big difference between their fate and yours? They have already become victims of this threat, while it spares you for a day or two. Unite, all of you, in your misfortune or in your peril! Defend what you still have, and reconquer what you have lost!

Otherwise your fate will be horrible, for we are in an age of science and method, and our rulers, served by an army of chemists and professors, are preparing a social structure for you in which all will be regulated as in a factory. There, the machine controls everything, even men, who are simple cogs to be disposed of when they take it upon themselves to reason and to will.

This is what has happened in the vast stretches of the great American West. Groups of speculators who are on very good terms with the government (as are all the rich, including those lucky scoundrels who have become rich) have been granted vast domains in fertile regions, which through large infusions of manpower and capital they have turned into grain factories. Here, farming takes place on the scale of an entire province. This vast space is entrusted to a sort of general, who is educated, experienced, good at farming and business, and skilled in the art of evaluating the exact value of the productive power of land and muscle. Our man establishes himself in a comfortable house in the center of his land. In his barns he has a hundred plows, a hundred sowers, a hundred harvesters, and twenty threshing machines. About fifty freight cars pulled by locomotives continuously come and go on the railways between the depots in the fields and the nearest port, whose piers and ships also belong to him. A network of telephones connects the palatial house to all the outbuildings of the estate. The master's voice is heard everywhere. He can listen to every sound and observe every act. Nothing is done without his orders, and nothing escapes his surveillance.

And what becomes of the worker or the peasant in this world that is so well organized? Machines, horses, and men are used in the same manner: they are viewed as so much force to be quantified numerically, and they must be used most profitably for the employer, with the greatest productivity and the least expense possible. The stables are laid out in such a way that as soon as the animals leave the building, they begin to plow the furrow several kilometers long that will extend to the end of the field. Each of their steps is calculated and each profits the master. Similarly, all of the workers' movements are regulated from the moment they leave the communal dormitories. There, neither women nor children come to disrupt the work with a hug or a kiss. The workers are grouped into squads, each with their sergeants, captains, and the inevitable informer. Their duty is to perform methodically the work they are ordered to do, and to observe silence in the ranks. When a machine breaks down, they throw it into the scrap heap if it is not possible to repair it. When a horse falls and breaks a leg, they shoot it in the ear and drag it to the mass grave. When a man succumbs to pain, breaks an arm or leg, or is incapacitated by fever, they are kind enough not to finish him off, but they get rid of him all the same:

they let him die out of the way, without annoying anyone with his moans. At the end of the vast undertaking, when nature takes a rest, the manager also takes a rest and fires his army. The following year, he will always find an adequate supply of muscle and bone to hire, but he will be careful not to employ the same workers as the previous year. They might speak of their experience, think that they know as much as the master, obey orders grudgingly, and who knows? They might become attached to the soil they cultivate and imagine that it belongs to them!

If the happiness of humanity consisted of creating a few millionaires who, to satisfy their whims and desires, hoarded the produce amassed by all the subjugated workers, then this scientific exploitation of the earth by an overseer of galley slaves would certainly be the ideal world. The financial results of these enterprises are extraordinary. The amount of grain yielded by the work of five hundred men can feed fifty thousand. The expense incurred for a meager wage yields an enormous amount of produce that is dispatched by the shipload and sold for ten times the cost of production. Even so, if the mass of consumers becomes too poor because they are without work and wages, they will be unable to buy all of this produce. Condemned to die of starvation, they will no longer enrich the speculators. But the latter are not in the least interested in the distant future. Their attitude is first to make money, travel the road paved with gold, and let the future take care of itself—the children will manage! *"Après nous le déluge!"*

This is your fate, fellow workers who love the plowed field where you saw for the first time the mystery of little sprouts of wheat piercing the hard lumps of soil. This is the fate they are preparing for you! They will take field and harvest away from you; indeed, they will take you yourselves and attach you to a screeching, smoking iron machine. There, completely surrounded by coal smoke, you will have to swing a lever back and forth ten to twelve thousand times every day. This is what they will call agriculture. In addition, you must not expect to go courting just because your heart tells you to take a wife. And don't even turn your head to look at the young woman who is passing by—the foreman will not like it if the boss is cheated of his labor. If it suits the boss to let you get married and have offspring, it is because he finds you to his liking. For you have the kind of slavish soul that he wishes most to mold, and you are abject enough for him to authorize you to help perpetuate the race of abjection. The future that has been awaiting you is that of the working man, woman and child of the factory! Never did ancient slavery more methodically mold and shape human material to reduce it to being a tool. What remnants of humanity are left in a haggard, twisted, scrofulous being who suffocates in an atmosphere thick with suint, grease, and dust?

Avoid this death at all costs, comrades. If you have a little plot of land, guard it jealously: it is your life and that of your wife and children whom you love. Join with your companions whose land, like yours, is threatened by factory owners, sport hunters, and money lenders. Forget all your little grudges against your neighbors and band together in communes where there is a solidarity of interests and where each clump of earth is defended by all members of the community. If you number a hundred, a thousand, or ten thousand, you will be quite strong against the lord and his valets, but you will not yet be strong enough to take on an army. Therefore, each commune must form an alliance with the others, so that the weakest might partake of the strength of all. In addition, you must call out to those who have nothing, to those disinherited people of the cities. You have perhaps been taught to despise them, but you must love them because they will help you to keep your land and to reconquer what has been taken from you. With them, you will attack and knock down the high walls of the enclosures. With them, you will found the great commune of men, where you will work in unison to invigorate and beautify the land, and live happily on this good earth that gives us bread.

But if you do not do this, all will be lost. You will die slaves and beggars. As the mayor of Algiers recently said to a delegation of humble, unemployed people: "So, you're hungry? Why don't you just eat one another!"

10

Anarchy (1894)

The following text was originally a talk presented on June 18, 1894, in Brussels to the members of "The Philanthropic Friends," a Masonic lodge. It was published as "L'Anarchie" in *Les Temps nouveaux* 18 (May 25–June 1, 1895).

Anarchy is far from being a new theory. The word itself, in its accepted meaning of "the absence of government" and "a society without leaders," is of ancient origin and was used long before the time of Proudhon.[1]

Besides, what difference do words make? There were "acratists" before there were anarchists, but the acratists were not given their name—a learned construction—until many generations had passed. In all ages there have been free men, those contemptuous of the law, men living without any master and in accordance with the primordial law of their own existence and their own thought. Even in the earliest ages we find everywhere tribes made up of men managing their own affairs as they wish, without any externally imposed law, having no rule of behavior other than "their own volition and free will," as Rabelais expresses it,[2] and impelled by their desire to found a "profound faith" like those "gallant knights" and "charming ladies" who gathered together in the Abbey of Thélème.

But if anarchy is as old as humanity, those who represent it nevertheless bring something new to the world. They have a keen awareness of the goal to be attained, and from all corners of the earth they join together to pursue their ideal of the eradication of every form of government. The dream of worldwide freedom is no longer a purely philosophical or literary utopia, as it was for the creators of the Cities of the Sun and the New Jerusalems.[3] It has become a practical goal that is actively pursued by masses of people united in their resolute quest for the birth of a society in which there are no more masters, no more official custodians of public morals, no more jailers, torturers and executioners, no more rich or poor. Instead there will be only brothers who have their share of daily bread, who have equal rights, and who coexist in peace and heartfelt unity that comes not out of obedience to law, which is always accompanied by dreadful threats, but rather from mutual respect for the interest of all, and from the scientific study of natural laws.

No doubt this ideal will appear chimerical to some of you, but I am sure that it will also seem desirable to most, and that you can see in the distance the ethereal image of a peaceful society in which men, henceforth reconciled with one another, will let their swords go to rust, melt down their cannons, and disarm their ships. Besides, aren't you among those who have long (for thousands of years, you say) worked to build the temple of equality? You are "masons," and the goal of your masonry is to construct an edifice of perfect proportions into which will enter only those who are free, equal and fraternal, who work ceaselessly to improve themselves, and in whom the power of love awakens a new life of justice and goodness. Isn't this your goal? And is it not true that there are others who share it? You claim no monopoly on the spirit of progress and renewal. Indeed, you do not commit the injustice of forgetting your express enemies who curse you and excommunicate you, the rabid Catholics who condemn to hell the enemies of the Holy Church but who themselves no less than you prophesy the coming of an age of lasting peace. Francis of Assisi, Catherine of Sienna, Theresa of Avila, and so many other adherents of a faith not at all your own certainly loved humanity with a most sincere love, and we must count them among those who lived for an ideal of universal happiness. And today there are millions upon millions of socialists, regardless of the school to which they belong, who also struggle for a future in which the power of capital will be broken, and in which men will finally be able to call themselves "equals" without irony.

The anarchists thus have a final goal in common with many other magnanimous persons belonging to a great diversity of religions, sects, and parties. But they distinguish themselves sharply from the others by their means, as their name indicates in the clearest terms. The conquest of power has almost always been the great preoccupation of revolutionaries, including the best intentioned of them. The prevailing system of education does not allow them to imagine a free society operating without a conventional government, and as soon as they have overthrown their hated masters, they hasten to replace them with new ones who are destined, according to the ancient maxim, to "make the people happy." Generally, no one has dared to prepare for a change of princes or dynasties without having paid homage or pledged obedience to some future sovereign. "The king is dead! Long live the king!" cried the eternally loyal subjects—even as they revolted. For many centuries this has been the unvarying course of history. "How could one possibly live without masters!" said the slaves, the spouses, the children, and the workers of the cities and countryside as they quite deliberately placed their shoulders under the yoke, like the ox that pulls the plow. One is reminded of the insurgents of 1830 who proclaimed "the best of republics"[4] embodied in the person of a new king, and the republicans of 1848 quietly repairing to their hovels after having undergone "three months of misery in the service of the provisional government."[5] During the same period, a revolution broke out in Germany and a popular assembly met in Frankfurt: "the old authority is a corpse," proclaimed one of the representatives. "Yes," replied the chairman, "but we will revive it. We will summon new men who know how to exercise power to restore the strength of the nation." On this topic it might be appropriate to repeat the line from Victor Hugo: "There is an age-old human instinct that leads to turpitude."[6]

In contrast to this instinct, anarchy truly represents a new spirit. One can in no way reproach the libertarians for seeking to get rid of a government only to put themselves in its place. "Get out of the way to make room for me!" are words that they would be appalled to speak. They would condemn to shame and contempt, or at least to pity, anyone who, stung by the tarantula of power, aspires to an office under the pretext of "making his fellow citizens happy." Anarchists contend that the state and all that it implies are not any kind of pure essence, much less a philosophical abstraction, but rather a collection of individuals placed in a specific milieu and subjected to its influence. Those individuals are raised up above their fellow citizens in dignity, power, and preferential treatment, and are consequently compelled to think themselves superior to the common people. Yet in reality the multitude of temptations besetting them almost inevitably leads them to fall below the general level.

This is what we constantly repeat to our brothers—including our fraternal enemies, the state socialists—"Watch out for your leaders and representatives!" Like you they are surely motivated by the best of intentions. They fervently desire the abolition of private property and of the tyrannical state. But new relationships and conditions change them little by little. Their morality changes along with their self-interest, and, thinking themselves eternally loyal to the cause and to their constituents, they inevitably become disloyal. As repositories of power they will also make use of the instruments of power: the army, moralizers, judges, police, and informers. More than three thousand years ago the Hindu poet of the Mahabharata expressed the wisdom of the centuries on this subject: "He who rides in a chariot will never be the friend of the one who goes on foot!"

Thus anarchists have the firmest principles in this area. In their view, the conquest of power can only serve to prolong the duration of the enslavement that accompanies it. So it is not without reason that even though the term "anarchists" ultimately has only a negative connotation, it remains the one by which we are universally known. One might label us "libertarians," as many among us willingly call themselves, or even "harmonists," since we see agreement based on free will as the constituting element of the future society. But these designations fail to distinguish us adequately from socialists. It is in fact our struggle against all official power that distinguishes us most essentially. Each individuality seems to us to be the center of the universe and each has the same right to its integral development, without interference from any power that supervises, reprimands or castigates it.

So you understand our ideal. The next question that arises is the following: "Is this truly a noble ideal? Does it justify the sacrifice of dedicated men and all the terrible risks that revolutions inevitably bring in their wake? Is anarchist morality pure, and if a libertarian society is created, will man be better off than in one based on fear of power and of the law? I reply with complete confidence (and I hope that soon you will join me in this response), "Yes, it is anarchist morality that is most in accord with the modern conception of justice and goodness."

The foundation of the old morality was, as you know, nothing but fear, that "trembling" of which the Bible speaks, and which was instilled in you through various teachings during your youth. "The fear of God is the beginning of wisdom" was formerly the starting point for all education. In other words, society as a whole is founded on terror. Men were not citizens, but rather subjects or members of a flock. Wives were servants and children slaves over whom the parents retained vestiges of the ancient law of life and death. We find everywhere, in all social relations, positions of superiority and subordination. In short, even in our own time the guiding principle of the state itself and of all the particular states that make it up is hierarchy, by which is meant "holy" archy or "sacred" authority, for that is the true meaning of the word.[7] This sacrosanct system of domination encompasses a long succession of superimposed classes in which the highest have the right to command and the lowest have the duty to obey. The official morality consists in bowing humbly to one's superiors and in proudly holding up one's head before one's subordinates. Each person must have, like Janus, two faces, with two smiles: one flattering, solicitous, and even servile, and the other haughty and nobly condescending. The principle of authority (which is the proper name for this phenomenon) demands that the superior should never give the impression of being wrong, and that in every verbal exchange he should have the last word. But above all, his orders must be carried out. That simplifies everything: there is no more need for quibbling, explanations, hesitations, discussions, or misgivings. Things move along all by themselves, for better or worse. And if a master isn't around to command in person, one has ready-made formulas—orders, decrees, or laws handed down from absolute masters and legislators at various levels. These formulas substitute for direct orders, and one can follow them without having to consider whether they are in accord with the inner voice of one's conscience.

Between equals, the task is more difficult but also more exalted. We must search fiercely for the truth, discover our own personal duty, learn to know ourselves, engage continually in our own education, and act in ways that respect the rights and interests of our comrades. Only then can one become a truly moral being and awaken to a feeling of responsibility. Morality is not a command to which one submits, a word that one repeats, something purely external to the individual. It must become a part of one's being, the very product of one's life. This is the way that we anarchists understand morality. Are we not justified in comparing this conception favorably with the one bequeathed to us by our ancestors?

Perhaps you will now concede that we are right. But here again, some of you will speak of a "chimera." Though it pleases me that you will at least concede that ours is a noble dream, I wish to claim more than this and assert that our ideal, our conception of morality, is fully in accord with the logic of history, brought about naturally through the evolution of humanity.

Long ago, haunted by their fear of the unknown as well as by their feeling of powerlessness to discover the real causes of things, men created out of their intense desire one or more helpful divinities that represented both their formless ideal and the basis for an entire mysterious world of things, both visible and invisible, that surrounded them. These phantoms of the imagination, invested with supreme power, also became in the eyes of men the principle of all justice and authority. These masters of the heavens needed interpreters on earth—magicians, counselors, and war chiefs before whom one learned to prostrate oneself as if before emissaries from on high. This was quite logical; however, man endures longer than his own works. For this reason the gods that he created never stop changing, like shadows cast into infinity. At first visible and driven by violent and fearsome human passions, they retreated little by little into an immense distance. Finally they became abstractions, sublime ideas that were no longer even assigned a name, and then merged with the natural laws that govern the world. They once again became part of a universe that they supposedly caused to burst forth out of nothingness. And now man finds himself alone on the earth, above which he once erected the colossal image of God.

Our entire conception of things changes simultaneously. If God disappears, those who derived from God their right to demand obedience will see their borrowed luster become tarnished. They will be obliged to return gradually to the ranks, adapting as best they can to the way things are. Today one can no longer find a Tamerlane,[8] who commanded his forty courtesans to jump from the top of a tower, certain that in the blink of an eye he would observe from the crenels forty bloody, broken corpses. Freedom of thought has made all men anarchists without their knowing it. Who today does not set aside a small corner of the brain for reflection? This is precisely the crime of crimes, the sin par excellence, symbolized by the fruit of the tree that revealed to men the knowledge of good and evil. From this came the hatred of science that the Church has always professed. From this came the rage that Napoleon, a modern Tamerlane, always harbored against the "ideologues."

But the ideologues have come. They have blown away the misty illusions of the past, undertaking once again the work of science through observation and experimentation. One of them, a nihilist before his time, an anarchist in word if not in deed, commenced by making a "tabula rasa" of all that he had learned.[9] Today there is scarcely a single scientific or literary scholar who does not claim to be his own master and model, the thinker of his own original thoughts, the moralist for his own morality. As Goethe said, "If you wish to blossom, blossom on your own." Do artists not seek to render nature as they see it, feel it, and understand it? One might see this as a kind of "aristocratic anarchy" that demands freedom only for the chosen people who consort with the muses,[10] those who climb Parnassus. Each of them wants to have freedom of thought and pursue his own ideal, without any limits, just as he pleases—while at the very same time saying that there must be "religion for the people." He wants to live as a fully independent man, but thinks that "obedience is designed for women." He wants to create original works of art, but thinks that "the crowd below" must remain in debasing, machine-like subservience to the operations of the division of labor! In any case, these aristocrats of taste and thought are powerless to close the floodgates against the coming deluge. If science, literature, and art have become anarchistic, if all progress and every new form of beauty are the result of the flourishing of free thought, this thought must also be at work within the depths of society. Today it is no longer possible to contain it. It is too late to stop the flood.

Isn't the loss of respect a quality par excellence of contemporary society? Some time ago I saw a crowd of thousands rush forward to gaze upon the empty carriage of a great lord. I no longer see such things. In India, pariahs once came devoutly to a halt the prescribed 115 paces from the haughty Brahmin. Since people began crowding into train stations, nothing has separated them but a partition in the waiting room. There are still more than enough examples of baseness and vile groveling in the world, but there has nevertheless been progress in the direction of equality. Before showing respect, one sometimes asks whether the man or the institution in question is truly respectable. One now considers the value of individuals and the importance of their deeds. Faith in greatness has disappeared, and when that faith no longer exists, the institutions that depended on it will in turn disappear. The abolition of the state is a natural implication of the dying out of such respect.

The anti-authoritarian critique to which the state is subjected applies equally to all social institutions. The people no longer believe in the sacred origin of private property, produced, as the economists have told us (though one doesn't dare repeat it today), by the personal labor of the property owners. They are well aware of the fact that the toil of one individual could never create by itself a fortune of millions upon millions, and that such a monstrous accumulation of wealth is always the result of defective social conditions, in which the product of the labor of thousands is allocated to a single person. They will always respect the hard-earned bread of the worker, the hut that he builds with his own hands, and the garden that he plants, but they are certainly going to lose their respect for the multitude of artificial holdings symbolized by various pieces of paper locked up in bank vaults. I have no doubt that the day will come when they will calmly reclaim possession of all the products of their common labor: mines and estates, factories and castles, railroads, ships, and cargo. When these masses, debased by their ignorance and the weakness that it inevitably produces, no longer deserve the terms with which they are insulted, when they come to know with complete certainty that the monopolization of these immense assets rests solely on fictitious scribbling and the sanctity of red tape, the prevailing social order will indeed be in danger! Considering the deep and irresistible evolution occurring in all human minds, the fanatical railing now directed against the innovators will seem so inane and devoid of sense to our descendents. What matters the filth spewed out by a press that has to pay back in choice prose the stipends of its patrons? What matters even the abuse heaped upon us quite sincerely by the "saintly but simple" religionists who would have gladly carried wood to burn John Huss at the stake![11] The movement that enthralls us is not the work of dull-witted troublemakers or pathetic dreamers, but that of the whole of society. It is necessitated by the progression of thought, which has now become as inevitable and ineluctable as the rotation of the heavens and earth.

Nevertheless, some doubt may remain in your minds whether anarchy has ever been any more than a mere ideal, an intellectual exercise, or the subject of dialectic. You may wonder whether it has ever been realized concretely, or whether any spontaneous organization has ever sprung forth, putting into practice the power of comrades working together freely, without the command of any master. But such doubts can easily be laid to rest. Yes, libertarian organizations have always existed. Yes, they constantly arise once again, each year in greater numbers, as a result of advances in individual initiative. To begin with, I could cite diverse tribal peoples called "savages," who even in our own day live in perfect social harmony, needing neither rulers nor laws, prisons nor police. But I will not stress such examples, despite their significance. I fear that some might object that these primitive societies lack complexity in comparison to the infinitely complicated organism of our modern world. Let us therefore set aside these primitive tribes and focus entirely on fully constituted nations that possess developed political and social systems.

Granted, I am unable to point to a single one throughout the course of history that has been constituted in a purely anarchistic manner, for each found itself in a period of struggle between diverse elements that had not yet been joined together with one another. But one finds that each of these fragmented societies, though not yet merged into a harmonious totality, was all the more prosperous and all the more creative to the degree that it had expanded freedom and accorded greater recognition to the value of each individual as a person. Since the point at which human society emerged from prehistory, awakened to the arts, sciences, and industry, and was able to hand down its experience to us through written records, the greatest periods in the lives of nations have always been those in which men, shaken by revolution, have suffered least under the long-lasting and heavy burden of a duly-constituted government. Judged by the progress in discovery, the flowering of thought, and the beauty of their art, the two greatest epochs for humanity were both tumultuous epochs, ages of "imperiled liberty." Order reigned over the immense empires of the Medes and the Persians, but nothing great came out of it. On the other hand, while republican Greece was in a constant state of unrest, shaken by continual upheavals, it gave birth to the founders of all that we think exalted and noble in modern civilization. It is impossible for us to engage in thought or to produce any work of art without recalling those free Hellenes who were our precursors and who remain our models. Two thousand years later, after an age of darkness and tyranny that seemed incapable of ever coming to an end, Italy, Flanders, and the Europe of the Free Cities reawakened. Countless revolutions shook the world. Ferrari[12] counted no less than seven thousand upheavals for Italy alone. In addition, the fire of free thought burst forth and humanity began once again to flourish. In the works of Raphaël, da Vinci, and Michelangelo it felt the vigor of youth once more.

Then came the great century of the Encyclopedists, with its proclamation of the rights of man and the world revolutions that ensued. One is hardly capable of listing all the advances that have been achieved since this great upheaval of humanity. It almost seems as if the greater part of all human history has been concentrated in this last century. The human population has increased to over half a billion. Commerce has increased more than tenfold. Industry has been transformed. The art of modifying natural resources has been wonderfully enriched. New sciences have appeared on the scene, and regardless of one's assessment of it, a third period in the history of art has begun. A conscious, worldwide socialist movement has begun to flourish. At the very least, one has the feeling of living in the century of great problems and enormous struggles. Imagine the hundred years that came in the wake of eighteenth-century philosophy being replaced with a period without history, such as that in which 400 million peaceful Chinese lived under the tutelage of a "father of the people," a ritualistic court, and mandarins armed with diplomas. Far from living with the great vigor that we have seen, we would have gradually fallen into a condition of inertia and death. Galileo, while locked away in the prisons of the Inquisition, could only murmur secretly, "Still, it moves!" But thanks to the revolutions and the fury of free thought, we can today cry from the housetops and in the public squares, "The world moves, and it will continue to move!"

In addition to this great movement that gradually transforms all of society in the direction of free thought, free morality and freedom of action, in short, toward the essentials of anarchy, there has also existed a history of direct social experimentation that has manifested itself in the founding of libertarian and communitarian colonies. These might be looked upon as a series of small tests that are analogous to the laboratory experiments of chemists and engineers. These efforts to create model communities all have the major failing of being created outside the normal conditions of life, that is to say, far from the cities where people intermingle, where ideas spring up, and where intellects are reinvigorated. Nevertheless, one can point to quite a number of such projects that have fully succeeded, for example, "Young Icaria," which is a transformation of Cabet's colony, founded almost half a century ago on principles of authoritarian socialism.[13] After repeated migrations, the group of communards has now become entirely anarchist and leads a simple life in the countryside of Iowa, near the Des Moines River.

But where anarchist practice really triumphs is in the course of everyday life among common people who would not be able to endure their dreadful struggle for existence if they did not engage in spontaneous mutual aid, putting aside differences and conflicts of interest. When one of them falls ill, other poor people take in his children, feeding them, sharing the meager sustenance of the week, seeking to make ends meet by doubling their hours of work. A sort of communism is instituted among neighbors through lending, in which there is a constant coming and going of household implements and provisions. Poverty unites the unfortunate in a fraternal league. Together they are hungry; together they are satisfied. Anarchist morality and practice are the rule even in bourgeois gatherings where they might seem to be entirely absent. Imagine a party in the countryside at which some participant, whether the host or one of the guests, would put on airs of superiority, order people around, or impose his whims rudely on everyone! Wouldn't this completely destroy all the pleasure and joy of the occasion? True geniality can only exist between those who are free and equal, between those who can enjoy themselves in whatever way suits them best, in separate groups if they wish, or drawing closer to one another and intermingling as they please, for the hours spent in this way are the most agreeable ones.

Please permit me at this point to relate to you a personal experience. We were sailing along in one of those modern ships that cleave through the waves at a speed of fifteen to twenty knots and trace a direct path from continent to content regardless of wind and tide. The air was calm, the evening pleasant, and the stars sparkled one by one in a black sky. We were conversing on the poop deck, and came inevitably to the eternal social question, which grabs us and seizes us by the throat like the Riddle of the Sphinx. The reactionary of the crowd was assailed by his interlocutors, who were all socialists, more or less. He turned suddenly toward the captain, our leader and master, hoping to find in him a born champion of the conventional wisdom. "You are the commander here! Isn't your authority sacred? What would happen to this vessel if it were not under the constant direction of your will?" "How naïve you are," replied the captain. "Just between us, I can tell you that for the most part I'm completely useless. The man at the helm keeps the ship on course, and in a little while another pilot takes his place, and later, still others, and we consistently follow our charted route, without my intervention. Below, the stokers and engineers do their work without my help or opinion, and do it better the less I meddle and give advice. And all the topmen are sailors who also know the jobs they have to do, and only occasionally do I have to coordinate my small part of the work with theirs, which is harder and less lucrative than mine. To be sure, I am charged with guiding the ship. But isn't it obvious that this is pure fiction? We use maps, but I didn't draw them. The compass guides us, but I didn't invent it. Someone dredged the channel of the port we left and the channel of the port to which we are heading. And this superb vessel, its ribs hardly groaning at all under the pressure of the waves, majestically rocking in the swell, powerfully steaming ahead—I didn't build it. What am I compared to the great men of the past, the scientists and inventors, our predecessors who taught us how to cross the seas? We are all their partners, including my comrades the sailors, and also you the passengers. After all, it is for you that we ride the waves, and in case of peril, we count on you to assist us fraternally. We have a common endeavor and we are united with one another!" Then everyone became silent, and I added to the treasury of my memory the precious words of this captain, the likes of whom is rarely encountered.

And so this vessel, this floating world in which, moreover, punishment is unknown, carries a model republic across the oceans, despite all the hierarchical labyrinths of the world. And this is hardly an isolated example. Each of you knows, if only by hearsay, of schools in which the professor, disregarding harsh regulations, treats all the pupils as friends and cordial colleagues. Everything required to get the little rascals under control is provided by the proper authorities, but their big friend has no need for all that paraphernalia of repression. He treats the children like human beings, constantly appealing to their good will, to their understanding of things, and to their sense of justice. And they all respond joyfully. A miniscule society that is anarchistic and truly humane is thus created, even though everything in the larger world seems to be in league to prevent its being born—laws, regulations, bad examples, and public immorality.

Anarchistic groups thus spring up constantly, despite all the old prejudices and the heavy weight of ancient customs. Our new world springs up all around us, like new flora sprouting up from the refuse of the ages. Not only is it not a mere dream, as some often claim, but it already manifests itself in a thousand different forms. One has to be blind not to notice it. On the other hand, if there is any form of society that is illusory and impossible, it is the pandemonium in which we now live. I hope that you will grant me that I have not gone overboard in my critique, though it is not difficult to do so in regard to the world we live in, which has given us the so-called principle of authority and the cut-throat struggle for survival. But in the end, by its very definition a society is a collection of individuals who come together and deliberate in pursuit of the common good. However, one cannot state unambiguously that the chaotic mass that we find around us constitutes a society. According to its proponents—and every bad cause possesses them—the goal of our society is supposed to be the attainment of perfect order through the satisfaction of the interests of all. But isn't it ludicrous to look for a well-ordered society anywhere in the sphere of European civilization, with its unending succession of internal conflict, murder and suicide, violence and shootings, depression and famine, theft, fraud, and deception of every kind, bankruptcy, collapse and ruin? Which of you, on leaving, will not see rising up around you specters of hunger and vice? In our own Europe there are five million men who are but waiting for the signal to kill other human beings, to burn houses and harvests. Ten million others on reserve outside the barracks are consciously committed to carrying out the same work of destruction. Five million wretches live, or at least vegetate, in prison, condemned to a variety of sentences; ten million die prematurely each year; and of 370 million persons, 350 million, that is to say nearly the whole, shudder in justifiable fear of the future. Despite the immense wealth of society, who among us could deny that an abrupt reversal of fortunes might take away his assets? These are the facts that no one can deny, and which ought I think inspire in us all the firm resolve to change the present state of things, which is ripe for permanent revolution.

I once had the opportunity to converse with a high-level bureaucrat, well trained by the daily routine of enacting laws and imposing penalties. "Go ahead, defend your society," I said to him. "How do you expect me to defend it?" he replied. "It is indefensible." Nevertheless it is defended, but with arguments that need no justification: with flogging, solitary confinement and the scaffold.

On the other hand, those who attack this society can do so with a completely clear conscience. It is certain that the movement of social transformation will involve violence and revolution, but isn't the world around us nothing but continual violence and permanent revolution? And between the two sides in the social struggle, which of the two sides will consist of responsible men? Those who proclaim an era of justice and equality for all, without distinction between classes and individuals, or those who wish to perpetuate the separation and consequently the hatred between castes, who add repressive law to repressive law, and whose only solution to social problems consists of infantry, cavalry, and artillery? History allows us to state with full confidence that a politics of hatred always begets more hatred, inevitably aggravates the overall situation, and may even bring on ultimate destruction. How many nations have gone to ruin in this way, oppressors as well as the oppressed! Will we in turn also go to ruin?

I hope that we will not, thanks to the anarchist thought that manifests itself ever more strongly, renewing human initiative. Aren't you yourselves, if not anarchists, then at least strongly tinged with anarchism? Who of you, in your heart of hearts, would call yourself the superior of your neighbor, and would not recognize in him your brother and your equal? The morality that has often been proclaimed here in words that are more or less symbolic will certainly become a reality. For we, as anarchists, know that this morality of perfect justice, liberty, and equality is surely the true one, and we live it with all our hearts, whereas our adversaries are uncertain. They are unsure of being right. At bottom, they are even convinced that they are wrong, and hand over the world to us in advance.

11

The Extended Family (1896)

This essay was published as "La grande famille" in *Le Magazine international* (January 1897). A previous English translation entitled "The Great Kinship" was made by the important but neglected libertarian theorist and cultural radical Edward Carpenter. See *Elisée and Elie Reclus: In Memoriam*, ed. Joseph Ishill (Berkeley Heights, N.J.: Oriole Press, 1927), 52–54.

Man likes to live in a dream world. The mental effort required to grasp reality seems too demanding, and he tries to avoid this struggle by resorting to ready-made opinions. If "doubt is the pillow of the wise," then blissful faith is the pillow of the simpleminded. Once there was a supreme God who did our thinking for us, willed and acted from on high, and guided human destiny according to his whims. His power was all that we needed, and it caused us to accept our inevitable fate with resignation or even gratitude. This personal God, on whom the meek could depend, is now perishing in his own temples, and men have to find a substitute for him. No longer do they have the All-Powerful at their service. They have only a few words that they try to endow with mysterious force or magical power—for example, the word "progress."

Without doubt, man has progressed in many ways. His sensations have become more refined, his thinking more acute and profound, and his humanity, embracing a much wider world, has expanded prodigiously. But no progress can occur without some degree of regression. The human being grows, but in the process he moves forward, thus losing part of the terrain that he formerly occupied. Ideally, civilized man should have kept the savage's strength, dexterity, coordination, natural good health, tranquility, simplicity of life, closeness to the beasts of the field, and harmonious relationship to the earth and all beings that inhabit it. But what was once the rule is now the exception. Much evidence suggests that a man with determination and a favorable environment can equal the savage in all his basic qualities, while also adding to them by means of a consciousness strengthened by a higher soul. But how many have gained without losing, who are the equals of both the primitive of the forest or prairie, and the modern artist or scholar in the bustling city?

And if sometimes a man of exceptional willpower and exemplary deeds manages to equal his ancestors in native qualities and even surpass them in acquired traits, one must still conclude that humanity as a whole has lost some of its early achievements. Thus the animal world, in which we find our origins and which instructed us in the art of existence, teaching us hunting and fishing, techniques of healing and of building houses, methods of working together and providing for our needs—this world has become more foreign to us. Whereas today we speak of the training or domestication of animals in the sense of subjugation, the primitive thought of his association with animals in fraternal terms. He saw in these living creatures his companions rather than his servants. Indeed, during times of danger, especially storms and floods, animals—dogs, birds, and snakes—took refuge with him.

The Indian woman of Brazil happily surrounds herself with a whole menagerie, and tapirs, deer, opossums, and even tame jaguars can be seen in the clearing around her cabin. Monkeys gambol in the branches above the hut, peccaries root in the ground, and toucans, hoccos,[1] and parrots perch here and there on the swaying branches, protected by dogs and large trumpeter birds.[2] And this entire republic functions without the need for an ill-tempered mistress to deal out insults or blows. The Quechuan[3] shepherd, crossing the plateau of the Andes with his llama and its packs, gains the cooperation of his beloved animal only through strokes and encouragement. If there were a single act of violence, the llama, his personal dignity offended, would lie down angrily and refuse to get up. He walks at his own pace, never accepts a burden that is too heavy, pauses for a long time at dawn to gaze upon the awakening sun, expects to be wreathed with flowers and ribbons or to have a banner waving above his head, and wants the women and children to pet and stroke him when he arrives at their huts. Isn't it also true that the horse of the Bedouin, another primitive, stays in the tent and the infants sleep between its legs?

The natural sympathy existing between all these beings brought them together in a pervasive atmosphere of peace and love. Birds perched on a man's hand, as they still do today on the horns of bulls, and squirrels frolicked within reach of the farmer or shepherd. Primitive people do not even exclude animals from their political communities. In Fazokl,[4] when the subjects depose their king, they always address the following speech to him: "Since you are no longer acceptable to men, women, children, and donkeys, the best thing you can do is to die, and we will help you do so."[5] Long ago, men and animals kept no secrets from one another. "The beasts talked," according to the fable, but more significantly, man understood. Are any stories more charming than those of southern India, which are perhaps the oldest traditional tales in the world, and which were passed on to the Dravidian invaders by the indigenous peoples? In these stories, elephants, jackals, tigers, lions, jerboas,[6] snakes, crayfish, monkeys, and men converse freely, thus forming, so to speak, the great common school of the primitive world. And in this school, the real teacher is usually the animal.

In these early periods, alliances between men and animals included a much greater number of species than are found today in the domestic sphere. Geoffroy St. Hilaire mentioned forty-seven of them that formed, so to speak, the entourage of man. But how many species that he failed to mention also lived long ago in intimacy with their youngest brother! He included neither the many companions of the Guarani[7] Indian woman, nor the snakes that the Dinka of the Nile[8] call by name and share their cows' milk with, nor the rhinoceroses that graze with the other livestock on the meadows of Assam,[9] nor the crocodiles of the Sindh,[10] which Hindu artists decorate with religious symbols. Archeologists have demonstrated conclusively that the Egyptians of the Old Kingdom[11] had among their herds of domestic animals three or even four species of antelope, and one ibex. After having been incorporated into human communities, all of these animals have now become wild again. Even hyena-like dogs and cheetahs were transformed by hunters into loyal companions. The Rig-Veda extols carrier pigeons, "swifter than the clouds." It sees in them gods and goddesses and directs that sacrifices be made and libations poured for them. And surely the mythical story of the Flood reminds us of our early ancestors' skillful use of the swift homing pigeon. Noah sent forth from the Ark a dove that explored the expanse of waters and the emerging land, and brought the olive branch back to him in its beak.

Today's domestication of animals exhibits in many ways moral regression since, far from improving animals, we have deformed and corrupted them. Although through selective breeding we have improved qualities such as strength, dexterity, scent, and speed in racing, as meat-eaters our major preoccupation has been to increase the bulk of meat and fat on four legs to provide walking storehouses of flesh that hobble from the manure pile to the slaughterhouse. Can we really say that the pig is superior to the wild boar or the timid sheep to the courageous mouflon?[12] The great art of breeders is to castrate their animals and create sterile hybrids. They train horses with the bit, whip, and spur, and then complain that the animals show no initiative. Even when they domesticate animals under the best possible conditions, they reduce their resistance to disease and ability to adapt to new environments, turning them into artificial beings incapable of living spontaneously in free nature.

Such degradation of species is itself a great evil, but civilized science goes even further and sets about exterminating them. We have seen how many birds have been wiped out by European hunters in New Zealand, Australia, Madagascar, and the polar archipelagos, and how many walruses and other cetaceans have already disappeared![13] The whale has fled the waters of the temperate zone, and soon will not even be found among the ice fields of the Arctic Ocean. All the large land animals are similarly threatened. We already know the fate of the aurochs[14] and the bison, and we can foresee that of the rhinoceros, the hippopotamus, and the elephant. Statistical estimates place the annual production of elephant ivory at eight hundred tons, which means that hunters kill forty thousand elephants, without counting those that are wounded and go deep into the bush to die. How far we have come from the Singhalese of long ago, for whom "the eighteenth science of man was to win the friendship of an elephant!" How far from the Aryans of India, who assigned to the tamed colossus two Brahmins as companions so that it might be taught to practice the virtues worthy of its breed!

On a plantation in Brazil, I once had the opportunity to observe the great contrast between the two modes of civilization. The owner took pride in two bulls that he had purchased at great expense in the Old World. One of them, which had come from Jersey, was pulling at a chain that passed through his nostrils, bellowing, fuming, pawing the ground with his hoof, pointing his horns, and looking menacingly at his keeper. The other, a zebu imported from India, followed us like a dog and with a sweet look begged to be petted. We poor, ignorant, "civilized" people, cooped up in our houses, distant from a nature that we dread because the sun is too hot or the wind too cold—we have even completely forgotten the meaning of the holidays that we celebrate. Though Christianity ignores the fact, all of them—Christmas, Easter, Rogation Days, and All Saints' Day—were originally nature festivals. Do we understand the meaning of the traditions that place the first man in a beautiful garden where he walked freely with all the animals, and that have the "Son of Man" born on a bed of straw, between the ass and the ox, the two companions of the plowman?

Although the distance that separates man from his animal brethren has widened, and though our direct interactions with species that remain free in the wilderness has diminished, it nevertheless seems clear that at least some progress has occurred, thanks to our closer relationship with domesticated animals not used for food. Without doubt, dogs have been to some degree corrupted. The majority are, like soldiers, accustomed to beatings and have become loathsome beings that tremble before the whip and cringe at the threatening words of the master. Others are trained to be vicious, like the bulldogs that bite the calves of poor people or go for the throats of slaves. Still others, those "greyhounds in jackets," take on all the vices of their mistresses—greediness, vanity, lust, and haughtiness. And the dogs in China, which are bred to be eaten, are of unsurpassed stupidity. But if a dog is truly loved and raised with kindness, gentleness, and nobility of feeling, doesn't it often realize a human or even superhuman level of devotion and moral goodness?

Cats have known much better than dogs how to maintain their personal independence and their distinctive character, so that we make alliances with them rather than taming them. Haven't they made almost miraculous intellectual and moral progress since they emerged from their original wild state in the forests? There is no human sentiment that they do not from time to time understand or share, no idea that they do not intuit, no desire that they do not anticipate. Poets have imagined them as magicians. And indeed, they do seem at times to be more discerning than their human friends in their foresightedness. And don't the "happy families" exhibited by showmen at fairs demonstrate that rats, mice, guinea pigs, and many other little creatures wish to attain, along with man, the great union of happiness and kindness? Every prison cell is soon transformed into a school for small animals—rats and mice, flies and fleas—provided the guards do not set things in order. The story of Pellisson's spider is well known.[15] The prisoner had regained his love of life, thanks to the little friend whose trainer he had become. But a guardian of order appeared, and with the avenging boot of official morality, crushed the creature that had come to console the poor wretch!

All of these facts demonstrate man's enormous resources for exerting a positive influence over the entire living world, which he now leaves to the mercy of fate and fails to connect to his own life. Some day our civilization, which is so fiercely individualist and divides the world into as many little belligerent states as there are private properties and family households, will finally collapse, and it will be necessary to practice mutual aid to assure our common survival. Some day the quest for friendship will replace the quest for material well-being that sooner or later will have been adequately provided for. Some day dedicated naturalists will have disclosed to us all that is charming, appealing, human, and often more than human in the nature of animals. We will then reflect upon all the species left behind in the march of progress and seek to make of them neither our servants nor our machines, but rather our true companions. Just as the study of primitive people has made a noteworthy contribution to our understanding of the civilized[16] man of our own time, so the study of the ways of animals will help us to delve more deeply into the life sciences, increase our knowledge of the nature of things, and expand our love. Let us look forward to the time when the deer emerges from the forest and, looking at us with its dark eyes, comes before us to be petted, and the bird, aware its own beauty, triumphantly perches on the shoulder of a beloved human companion, asking her for its share of love.

12

Evolution, Revolution, and the Anarchist Ideal (1898)

On February 5, 1880, Reclus delivered an address in Geneva entitled "Evolution et Révolution." It was published in the journal *Le Révolté* under that title (February 21, 1880): 1–2, and then was reprinted as a pamphlet and translated many times. Reclus finally expanded the discussion into a book entitled *L'Evolution, la révolution et l'idéal anarchique* (Paris: Stock, 1898; Montréal: Lux Editions, 2004), his only full-length work on anarchist politics. The following text consists of excerpts containing the most important discussions in that work; it includes about one-fourth of the original text.

Evolution encompasses the entirety of human affairs. We ought to recognize that revolution does also, even though this parallelism is not always evident from the individual events that make up the whole of the life of societies. All advancements are interdependent, and in proportion to our knowledge and power, we desire them all—social and political progress, moral and material progress, and progress in science, art, and industry. In every sphere we are not only evolutionists, but just as much revolutionists, since we realize that history itself is but a series of achievements that follows a series of preparations. The great intellectual evolution that emancipates minds has a logical consequence in the emancipation of individuals in all of their relationships with other individuals.

It can thus be said that evolution and revolution are two successive aspects of the same phenomenon, evolution preceding revolution, and revolution preceding a new evolution, which is in turn the mother of future revolutions. Can any change take place without producing sudden shifts in the balance of life? Does revolution not inevitably follow evolution in the same way that an act follows the will to act? The two differ only in the time of their appearance. When a mass of fallen debris obstructs a river, the waters gradually accumulate above the impediment, and a lake is formed through slow evolution. Then suddenly the down-river dam begins to leak, and the fall of a pebble precipitates a cataclysm. The obstruction is violently swept away, and the emptied lake once again becomes a river. In this way, a small terrestrial revolution takes place.

If revolution always lags behind evolution, it is because of the resistance of the environment: the water in a stream splashes between its banks because they hinder its flow; thunder rumbles in the sky because the atmosphere resists the electrical charge that flashes down from the cloud. Each transformation of matter and each realization of an idea is, during its actual process of change, thwarted by the inertia of the environment. A new phenomenon can thus come into being only through an effort that is more violent, or a force that is more powerful, than the resistance. Herder, speaking of the French Revolution, expressed this idea: "A seed falls to the ground, and for a long time it seems to be dead. Then suddenly it sprouts, displaces the hard soil that had covered it, violently pushes away its enemy, the clay, and thus becomes a flowering plant that bears fruit." And consider how a child is born: after spending nine months in the darkness of the womb, it also escapes violently, tearing its receptacle and sometimes even killing its mother. Such are revolutions—necessary consequences of the evolutions that preceded them.

However, revolutions do not necessarily constitute progress, just as evolutions are not always directed toward justice. Everything changes; everything in nature moves as part of an eternal movement. But where there is progress, there can also be regression, and if some evolutions tend toward the growth of life, there are others that incline toward death. To stop is impossible, and it is necessary to move in one direction or another. The hardened reactionary and the gentle liberal, both of whom cry out in fright at the word "revolution," nevertheless march onward toward a revolution—the last one, which is eternal rest. Disease, senility, and gangrene are evolutions just as much as puberty is. The appearance of worms in a corpse, like the first cry of an infant, indicates that a revolution has occurred. Physiology and history demonstrate that some evolutions indicate decline, and certain revolutions mean death.

We know human history only partially, based on the experience of but a few thousand years, yet it offers endless examples of tribes, peoples, cities, and empires that have perished miserably as a consequence of slow evolutions that led to their downfall. The factors that brought about these maladies of entire nations and races have been manifold and diverse. Climate and soil can deteriorate, as has certainly happened over vast stretches of

Central Asia where lakes and rivers have dried up and salt deposits have reclaimed previously fertile lands. Invasions of enemy hordes devastated certain regions to such an extent that they have remained forever desolate; however, many nations were able to flourish again following conquests and massacres, even after centuries of oppression. Thus if a nation falls again into barbarism or completely dies out, one must seek the reasons for its regression and ruin, above all within the nation itself and in its essential constitution, rather than in external circumstances. There is a fundamental cause—indeed, the cause of all causes—that epitomizes the history of decline. It is the establishment of mastery of one part of society over another, and the monopolizing of land, capital, power, education, and honors by a few or by an aristocracy. As soon as the dull masses no longer have the drive to revolt against this monopoly by a small number of men, they are as good as dead, and their disappearance is but a matter of time. The black plague will soon come to finish off such a useless swarm of individuals without liberty. Slaughtering invaders charge from east and west, and the desert moves in to replace immense cities. Thus Assyria and Egypt died and Persia collapsed, and when the whole Roman Empire belonged to a few great landowners, the barbarians soon replaced the enslaved proletariat.

Every event is two-sided, for it is at once a phenomenon of death and a phenomenon of revival; in other words, it is the result of evolution toward decay and also toward progress. Thus the Roman Empire's destruction, in its immense complexity, consisted of a whole set of revolutions corresponding to a series of evolutions, some of which were disastrous and others fortunate. The destruction of the formidable machine of suppression that had weighed heavily on the world was certainly a great relief for the oppressed, and the violent arrival of the peoples from the north to the world of civilization was also in many respects a fortunate stage in the history of humanity. During the upheaval, many of the enslaved regained a small amount of liberty at the expense of their masters; however, science and industry perished or went into hiding. Statues were smashed and libraries were burned. It seems as though the chain of time had broken, so to speak. The people abandoned their heritage of knowledge. Despotism was followed by a worse despotism, and from a dead religion grew the offshoots of a new one that was more authoritarian, cruel, and fanatical. For a thousand years, the darkness of ignorance and folly propagated by monks spread across the earth.

Since every event and every period of history presents a double aspect, it is impossible to judge any of them categorically. The very example of the renewal that brought the Middle Ages and the night of ignorance to a close shows us how two revolutions can simultaneously be carried out—one resulting in decline, and the other in progress. The Renaissance, which rediscovered the monuments of antiquity, deciphered its books and teachings, freed science from superstitious methods, and once again engaged men in objective studies, also resulted in the definitive end of the spontaneous artistic movement that had developed so splendidly during the period of free cities and communes. It came as suddenly as the overflowing of a river that ravages the neighboring farm lands. Everything had to start over, and often, banal imitations of the ancient replaced works that at least had the merit of being original!

The renaissance of science and art was accompanied in the religious world by the split within Christianity called the Reformation. For a long time, it seemed natural to view this revolution as one of the beneficial turning points of humanity, epitomized by the conquest of the right of individual initiative and the emancipation of the mind, which the priests had kept in servile ignorance. It was believed that henceforth, men would be their own masters, each equal through the independence of thought. But we now know that the Reformation also meant the establishment of other authoritarian churches in opposition to the one that had hitherto held the monopoly on intellectual enslavement. The Reformation shifted fortunes and prebends to benefit the new power, and religious orders emerged from both sides—Jesuits and counter-Jesuits—to exploit the people in new ways. Luther and Calvin spoke of those who did not share their views with the same language of fierce intolerance as such figures as St. Dominic and Innocent III. Like the Inquisition, they spied, imprisoned, quartered, and burned, and in principle, their doctrine implied equal obedience to kings and to the interpreters of the "Divine Word."

There is often a most shocking disparity between the revolutionary circumstances that accompany the emergence of an institution and the manner in which it functions, which is completely opposed to the ideals of its naïve founders. At its birth, there may be cries of "Liberty! Liberty!" and the hymn "War against the Tyrants" may resound in the streets; however, "tyrants" still manage to come into their midst as the direct result of the routine, the hierarchy, and the spirit of regression that gradually encroach on every institution. The longer any institution persists, the more formidable it becomes, for it finally rots the very soil on which it stands and pollutes the atmosphere around it. The mistakes that it sanctions, and the perversion of ideas and feelings that it justifies and promotes, take on such a character of antiquity and even sanctity that rarely does anyone dare to challenge it. Its authority grows with each passing century, and if it nevertheless dies out in the end, as do all things, it is because it finds itself increasingly at odds with the totality of new developments emerging around it.

Some institutions, such as those of religious creeds, have gained such a powerful hold over the soul that many free-thinking historians have thought it impossible for men to liberate themselves from them. Indeed, the popular image of God sitting on his throne in Heaven is not one that is easy to overcome. In the logical order of human development, religious organization followed the political one, and priests came after chiefs, since every image presupposes a primordial reality; however, the religious illusion was placed at the loftiest height in order to make it the original justification for all earthly authority, and it was thus endowed with a majestic character par excellence. One spoke to the sovereign and mysterious power, the "Unknown God," in a state of fear and trembling that suppressed all thought and all inclination toward critical analysis or personal judgment. Adoration was the only feeling that priests allowed their faithful.

According to social psychology, we must mistrust not only the power that is already established, but also that which is emerging. It is equally important to examine carefully the practical meaning of such seemingly innocuous or even seductive words as "patriotism," "order," and "social peace." The love of one's native soil is, without doubt, a very natural and agreeable sentiment. It is delightful for an exile to hear his cherished maternal language and once more to see places that remind him of his birthplace. And such love is not limited to the land that nourished him and the language that he heard in the cradle, but also extends, through a natural impulse, to the sons of that same land, with whom he shares ideas, feelings, and customs. If he has a noble nature, he will finally become fervently attached, with a passionate solidarity, to those whose needs and ways he knows intimately. If this is "patriotism," what man of heart could help but feel it? But the word almost always hides a meaning quite different from mere "love for the land of one's forefathers."

It is a strange contradiction that one's native land has never been spoken of with such burning affection as it has since it began to disappear gradually into the great terrestrial homeland of humanity. Flags are seen everywhere, especially at the doors of cabarets and houses of ill repute. The "ruling classes" incessantly boast of their patriotism, while at the same time investing their assets abroad and trading illicitly with Vienna or

Berlin in whatever will bring them money—including state secrets. Even scholars, forgetful of having once constituted an international republic throughout the world, speak of "French science," "German science," or "Italian science," as if it were possible to confine the knowledge of facts and the dissemination of ideas within borders, under police protection. They practice protectionism not only for turnips and cotton cloth, but also for the products of the mind. But to the degree that the minds of the powerful become narrower, those of ordinary people are expanded. Men in high positions see their domain and their hopes diminished to the extent that we rebels take possession of the universe and enlarge our hearts. We think of ourselves as comrades throughout the world, from America to Europe and from Europe to Australia. We use the same language to assert the same interests, and the day is coming when we will in a spontaneous impulse adopt the same tactics and a single rallying cry. Our army awakens in all corners of the world.

In comparison to this global movement, what is commonly called patriotism is nothing but a regression in every respect. One would have to be extremely naïve to be unaware of the fact that the "catechisms of citizenship" preach the love of homeland in order to serve all the interests and privileges of the ruling class, and that for the benefit of this class they promote hatred between the weak and disinherited of various countries. The term "patriotism" and all the modern glosses on it disguise the age-old practices of servile obedience to the will of a leader and the complete abdication of the individual in the face of those who hold power and wield the entire nation like a blind force. Similarly, the words "order" and "social peace" sound quite beautiful to our ears, but we would like to know what those noble apostles, the rulers, mean by these words. Yes, peace and order are great ideals that deserve to be realized, but under one condition: that this peace is not that of the grave, and this order is not that of Warsaw![1] Our future peace must arise not from the unquestioned domination by some and the hopeless enslavement of others, but from good, straightforward equality among friends.

Although the current state of affairs is atrocious, an immense evolution has taken place, giving promise of the next revolution. This evolution consists in the fact that the "science" of economics, which prophesied scarcity of resources and the inevitable death of the starving masses, has been proven wrong, and that moreover, a suffering humanity, believing itself to be poor only a short time ago, has discovered its wealth. Thus its ideal of "bread for all" has been found to be no mere utopia. The earth is vast enough to nourish us all and rich enough to support us comfortably. It can provide enough crops for all to have food, it produces enough fibrous plants for all to have clothing, and it contains enough stones and clay for all to have houses. This is economic reality in all its simplicity. Not only is that which the earth produces adequate for the consumption of its present inhabitants, but it would also be enough if consumption were suddenly to double. This would be the case even if science did not intervene to advance agriculture beyond its empirical methods by placing at its disposal all of the resources now available from chemistry, physics, meteorology, and mechanics. In the great family of humanity, hunger is not only the result of a collective crime, it is moreover an absurdity, since production is more than double what is needed for consumption.

❧

And what of freedom of speech and freedom of action? Are they not direct and logical consequences of freedom of thought? Speech is but thought become audible. Action is but thought become visible. Our ideal thus entails for each man the complete and absolute liberty to express his thoughts in every area, including science, politics, and morals, without any condition other than his respect for others. It also entails the right of each to do as he pleases while naturally joining his will with those of others in all collective endeavors. His own freedom is in no way limited by this union, but rather expands, thanks to the strength of the common will.

It goes without saying that this absolute freedom of thought, speech, and action is incompatible with the maintenance of institutions that restrict free thought, rigidify speech in the form of a final and irrevocable vow, and even dictate that the worker fold his arms and die of hunger at the owner's command.[2] Conservatives are by no means wrong when they generalize that revolutionaries are "enemies of religion, family, and property." Yes, anarchists reject the authority of dogma and the intrusion of the supernatural into our lives. In this sense, whatever fervor they may bring into the struggle for their ideals of fraternity and solidarity, they are enemies of religion. Yes, they want to abolish matrimonial trafficking and instead desire free unions based only on mutual affection, self-respect, and the dignity of others. In this sense, as loving and devoted as they are to those whose lives are joined with theirs, they are indeed enemies of the family. Yes, they want to abolish the monopolizing of the earth and its products in order to distribute them to everyone. In this sense, the happiness they would have in guaranteeing to everyone the enjoyment of the fruits of the earth makes them enemies of property. Certainly, we love peace, and our ideal is harmony among all men. Yet war rages around us. It appears before us in the distance as a sad prospect, for in the immense complexity of human affairs, the march toward peace is itself accompanied by struggles. "My kingdom is not of this world," said the Son of Man. Still, he also "brought a sword," creating "the division between son and father, and between daughter and mother." Every cause, even the worst, has its defenders, and even though the revolutionary loves them, he must nonetheless also fight against them.

❧

Nothing good can possibly come to us from the republic and the successful "republicans," that is, those who gain power. To hope otherwise would be to accept a historical absurdity, utter nonsense. The class that possesses and governs is inevitably the enemy of all progress. The vehicle of modern thought and of intellectual and moral evolution is that part of society which struggles, labors, and is oppressed. It is that part which develops and realizes the idea and which, with great difficulty, constantly sets the chariot of society in motion, while conservatives endlessly try to stall it or bog it down.

But one might ask whether our evolutionary and revolutionary friends, the socialists, are equally liable to betray their cause, and whether we will see them one day go through the usual process of regression after those among them who want to "conquer state power" have succeeded. If the socialists become our masters, they will certainly proceed in the same manner as their predecessors, the republicans. The laws of history will not bend in their favor. Once they have power, they will not fail to use it, if only under the illusion or pretense that this force will be rendered useless as all obstacles are swept away and all hostile elements destroyed. The world is full of such ambitious and naïve persons who live with the illusory hope of transforming society through their exceptional capacity to command; however, when they have risen into the ranks of the leaders, or at least become enmeshed in the vast machine of high-level administration, they understand that their isolated will has no hold over the only real power, which is the inner workings of public opinion, and that all their efforts risk being lost amidst the indifference and ill will that surrounds them. What remains for them to do but follow governmental routine, enrich their families, and dole out positions to their friends?

Some fervent authoritarian socialists tell us that the mirage of power and the exercise of authority can doubtless have grave dangers for men who are simply motivated by good intentions, but that this danger need not be feared by those who have mapped out their plan of action through a program rigorously debated with comrades who will know how to call them to order in case of negligence or betrayal. It is required that programs be duly spelled out, signed, and countersigned. They are published in thousands of copies. They are posted on the doors of meeting halls, and each candidate knows them by heart. Are these not sufficient guarantees? However, the meaning of these scrupulously debated words varies from year to year according to events and perspectives, and each understands it according to his own interests. And when a whole faction comes to view things differently than it had in the beginning, the clearest statements take on a merely symbolic meaning, and eventually become no more than historical documents.

The fact is that those who aspire to conquer state power must obviously use the means that seem to lead most surely to their goal. In republics with universal suffrage, they court the multitudes, the crowd. They support the interests of the wine industry and make themselves popular at the pubs. They welcome voters from wherever they come, unconcerned about sacrificing substance for form. They invite enemies into their midst, which is like injecting poison directly into the body. In countries with a monarchy, many socialists declare themselves to be indifferent to the form of government, and they even call upon the king's ministers to help them realize their plans for social change, as if it were logically possible to reconcile domination by a single ruler with brotherly mutual aid among men. But the impatience to act can blind one to obstacles, and faith willingly believes that it can move mountains. Lassalle longs to have Bismarck as a partner in establishing a new world.[3] Others turn toward the Pope, asking him to head the league of the humble. And when the young Emperor of Germany gathered a few philanthropists and sociologists at his table, there were those who imagined that the new day had finally dawned.

And if some socialists are still fascinated by the prestige of political power expressed as divine right or the right of force, they succumb even more readily to power that is masked by its popular origin in limited or universal suffrage. In order to win votes, or in other words, to earn the favor of the citizens, which initially seems quite legitimate, the socialist candidate readily flatters the tastes, the inclinations, or even the prejudices of his electorate. He blithely ignores disagreements, disputes, and grudges, and for a while becomes the friend, or at least the ally, of those with whom only a short time ago he had exchanged invectives. In the clericalist, he tries to find a Christian socialist. In the liberal bourgeois, he conjures up the reformer. And in the patriot, he appeals to the courageous defender of civic dignity. At times, he even takes care not to scare off the "owner" or the "boss." He goes so far as to present them with his demands as if they were guarantees of peace. "May Day," which was supposed to represent victory in a long struggle against Lord Capital, has become a holiday with garlands and farandoles. With these superficial gestures to the voters, the candidates gradually forget the proud language of truth and the uncompromising attitude of combat. Their very spirit undergoes a pervasive transformation, especially among those who reach the goal of all their efforts and assume their places on velvet benches facing the gold-fringed rostrum. At this point they must become experts at exchanging smiles, handshakes, and favors.

This is simply human nature, and it would be absurd on our part to hold a grudge against the socialist leaders who, finding themselves caught up in the electoral machine, end up being gradually transformed into nothing more than bourgeois with liberal ideas. They have placed themselves in determinate conditions that in turn determine them. The consequences are inevitable, and the historian should limit himself to pointing it out as a danger to revolutionaries who would rashly throw themselves into the political fray. Besides, one need not exaggerate the results of this evolution of socialist politicians, for the struggling masses are always composed of two elements whose respective interests must increasingly diverge. Some are bound to abandon the original cause, while others remain faithful to it. These developments imply a new categorization of individuals, in which they are grouped according to their actual affinities. Thus we have recently seen the Republican Party split in two, forming on the one hand the crowd of "opportunists," and on the other, the socialist factions. The latter will also have to divide, one group watering down its program to make it more palatable to the conservatives, and the other group maintaining its spirit of straightforward evolution and honest revolution. After having had their moments of discouragement and even skepticism, they will "let the dead bury the dead" and will return to take their place among the living. But if only they knew that every party requires *esprit de corps*, and, consequently, solidarity in evil as well as in good. Each member of a party becomes bound up with the mistakes, the lies, and the ambitions of all his comrades and masters. It is only the free man—who of his own accord joins his strength with that of other men acting out of their own will—who has the right to disavow the mistakes or misdeeds of his so-called companions. He takes responsibility only for himself.

❧

Since the present function of the state consists foremost of defending the interests of landowners and the "rights of capital," it is indispensable for the economist to have at his disposal some successful arguments and fantastic lies that the poor, wanting very much to support the national economy, can accept without question. But alas! These fine-sounding theories, invented in the past for consumption by the dull masses, can no longer be accepted. One might well blush to repeat the old assertion that "work is always rewarded by wealth and property." In claiming that labor is the source of wealth, the economists are perfectly conscious that they are not telling the truth. Like the socialists, they know that great wealth is not the product of personal effort, but of the work of others. They are not unaware that speculation and success in the stock market, the source of great fortunes, can be justly compared with the exploits of bandits. They certainly would not dare to claim that the individual who has a million to spend each week, which is equivalent to the amount necessary to support a hundred thousand persons, distinguishes himself from other men through intelligence and virtue a hundred thousand times superior to that of the average person. It would be foolish, and almost complicitous, to waste time debating the hypocritical arguments on which this alleged source of social inequality is based.

But another kind of reasoning is used that at least has the merit of not being founded on a lie. Against the demands made on behalf of society, some invoke the right of the strongest, and even the respected name of Darwin (though without regard for his actual views), in order to defend injustice and violence. The strength of muscle and jaw, of cudgel and bludgeon—this is the ultimate argument! In actuality it is the right of the strongest that triumphs in the monopoly of wealth. The one who is best equipped materially, the most favored through birth, education, and friends, the best armed through force or trickery, and who finds before him the weakest enemies, has the best chance of succeeding. He is more capable than others of building a lofty fortress from which he can fire on his unfortunate brothers.

Thus the crude battle of conflicting egos determines the outcome. But in former times, one hardly dared acknowledge this theory of iron and fire, which seemed too violent, and preferred instead the language of hypocritical virtue. It was veiled by solemn expressions so that the people would not understand the meaning. "Work is a bridle," said Guizot.[4] But naturalists' studies of the struggle between species for existence and the concept of the survival of the fittest have encouraged the theoreticians of force to announce brazenly their arrogant challenge. "You see," they say, "this is the inevitable law and immutable destiny to which both predator and prey are subject."

We should be pleased that the question is thus simplified in all its brutality, for it is that much closer to being resolved. "Force rules!" say the defenders of social inequality. "Yes, it's force that rules!" cry out ever more loudly those who profit from modern industry in its ruthless development, the desired result of which is above all to reduce the number of workers. But could not the revolutionaries say much the same thing as the economists and the industrialists, but with the understanding that cooperation for existence will gradually replace the struggle for survival? The law of the strongest will not always benefit the industrial monopoly. "Might makes right," said Bismarck, like many others before him, but the day is coming when might will be at the service of right. If it is true that ideas of solidarity are spreading, that the conquests of science will eventually reach every level of society, and that moral resources are becoming the property of all, will not the workers, who have both the right and the might, make use of these things in order to create a revolution for the benefit of all? As strong as they may be in money, intelligence, and shrewdness, what can isolated individuals do against the united masses? The rulers have lost hope of giving any moral justification for their cause; now they ask only to rule with a firm hand. This is the only superiority to which they aspire. One could easily cite examples of state officials who were chosen not for their military glory, their noble genealogy, their talents, or their eloquence, but solely for their lack of scruples. In this regard, one has full confidence in them: they allow no prejudice to stand in the way of the conquest of power or the defense of bank notes.

In no modern revolution have the privileged been known to fight their own battles. They always depend on armies of poor people, whom they indoctrinate with the so-called religion of the flag and drill in the so-called maintenance of order. Six million men, without counting all the ranks of police, are employed for such work in Europe. But these armies can disintegrate. They can recall the common origin and destiny that connect them with the masses of the people, and the hand that leads them can lose control. Composed largely of proletarians, they can and certainly will become for bourgeois society what the barbarians in the pay of the empire became for Roman society—an element of dissolution. History is full of examples of outbreaks of panic to which the powerful succumb, even those who have preserved their strength of character—for there are also a number of "rulers" who are nothing more than simple degenerates. The latter are the sort who, if trapped in a spreading fire, would not have the energy and physical strength (even if there were a hundred of them) to break through a wooden wall, nor enough dignity to allow women and children to escape first.[5] When the disinherited are united in their own interests—trade with trade, nation with nation, race with race, or, spontaneously, man with fellow man—and when they know their goal well, there can be no doubt that the opportunity will arise for them to use their force in the service of liberty for all. As powerful as the master may be at that time, he will be quite weak when he faces all who are united with a single will and who rise up against him so that they may be assured from that moment on of their bread and liberty.

Ignorance is decreasing, and among the united revolutionary evolutionists, knowledge will soon be the guide of power. This is the overriding fact that gives us confidence in the promise of humanity: despite the infinite complexity of things, history demonstrates that progress will win out over regression. In considering all the facts of contemporary life, some attest to a relative decline while others conclude that there has been a step forward. The latter view is more valid, since day by day evolution moves us ever closer to that totality of both peaceful and violent transformations that we already call "social revolution." This will entail above all the destruction of the despotic power of persons and things, and of individual monopoly over the products of collective labor.

The major event in this evolutionary process is the emergence of the Workers' International. No doubt it has been taking root ever since men of different nations began practicing mutual aid, in complete friendship and for their common interests. It even acquired a theoretical existence when the philosophers of the eighteenth century inspired the French Revolution's proclamation of the "rights of man." But these rights have remained a mere slogan, and the assembly that shouted them out to the world was careful not to put them into practice. It did not even dare to abolish the slavery of the blacks of Santo Domingo and only yielded after years of insurrection, when it seemed that the last chance for salvation was to pay this price. The International, which was in the process of formation in all civilized countries, did not fully come to consciousness of itself until the second half of our century, and it was in the sphere of labor that it emerged. The "ruling classes" had nothing to do with it. The International! Since the discovery of America and the circumnavigation of the earth, no achievement was more important in the history of man. Columbus, Magellan, and El Cano[6] were the first to notice the physical unity of the earth, but the future normative unity that the philosophers desired began to be realized only when the English, French, and German workers, forgetting their different origins and understanding one another in spite of their diversity of languages, joined together to form a single nation, in defiance of all their respective governments. The beginnings of the undertaking were not impressive. Scarcely a few thousand men banded together in this association, which was the original cell of the humanity of the future. But historians understood the fundamental importance of the event that had just occurred. In the early years of the International, the overturning of the Vendôme Column during the Paris Commune showed that the ideas of that organization had become a living reality. Until that time, it was unheard of for a conquered people to overturn enthusiastically the monument of former victories. It was done not to flatter in a cowardly manner those who had in turn just conquered them, but rather to show their fraternal sympathy toward the brothers who had been driven against them, and their feelings of loathing for the masters and kings who on both sides had led their subjects to the slaughterhouse. For those who know how to rise above the petty struggles of factions and contemplate the march of history from a distance, there has never in this century been a more impressive sign of the times than the toppling of the imperial column onto a pile of manure![7]

❧

As soon as the spirit of demanding their due pervades the entire mass of the oppressed, every event, even if it seems to be of minimal importance, will be capable of creating shock waves of change, just as a single spark can cause a whole keg of powder to explode. Already we see harbingers of the great struggle. For example, when in 1890 the call of "May Day" resounded, launched by an unknown person, perhaps an Australian comrade, the workers of the world were suddenly united in a single thought. On that day, the International, which had officially been buried, was brought back to life—not by the command of leaders, but through the pressure of the masses. Neither the "wise counsel" of influential socialists nor the oppressive apparatus of governments was able to prevent the oppressed of all nations from feeling that they are brothers throughout the world, and from affirming this to one another. However, on the surface, "May Day" did not seem to amount to much, merely a platonic expression, a rallying cry, a password! Bosses and governments, aided by the socialist leaders themselves, have indeed attempted to turn those fateful words into nothing more than an empty formula. Nevertheless, this cry and this yearly celebration have taken on an epic significance through their universality.

Another kind of outcry, one that is sudden, spontaneous, and unexpected, can lead to even more surprising results. Due to one cause or another and in relation to some insignificant fact, the force of circumstances—that is, the entirety of economic conditions—will inevitably give rise to the kind of crisis that impassions even the indifferent. At that moment, there will suddenly be an explosion of the tremendous energy that has accumulated in the hearts of men because of a violated sense of justice, unredressed sufferings, and unappeased hatreds. Any day might bring such a cataclysm. The firing of a worker, a local strike, or an unforeseen massacre can be the cause of revolution, for the feeling of solidarity is constantly spreading, and each local tremor tends to shake all of humanity. Several years ago, the new rallying cry of "general strike" burst forth in the factories. This term seemed bizarre, and was thought to express a mere dream or chimerical hope. But it was repeated more loudly, and now it resounds so strongly that the capitalist world often trembles from it. No, the general strike is not impossible. English, Belgian, French, German, American, and Australian wage workers understand that it is up to them to withhold all labor from their bosses on the same day. And why would they not carry out tomorrow what they understand today, especially if a soldiers' strike is added to that of the workers? The newspapers unanimously maintain an extremely careful silence when soldiers rebel or leave the military en masse. Conservatives, who prefer to ignore completely any facts that are not in accord with their wishes, would like to believe that such a social abomination is impossible. But collective desertions, partial rebellions, and refusals to shoot are phenomena that occur frequently in poorly trained armies, and that are not completely unknown in the toughest of military organizations. Those among us who remember the Commune are reminded of the thousands of men left in Paris by Thiers, who were disarmed by the people and easily converted to its cause. When the majority of soldiers are pervaded by the will to strike, the opportunity to act upon it will come sooner or later.

The strike, or rather the spirit of the strike in the broadest sense, derives its value above all through the solidarity that it creates among those who demand their rights. In struggling for a common cause, they learn to love one another. But there are also efforts at direct association, and these also contribute increasingly to the social revolution. This uniting of forces by the poor, the farmers, or the industrial workers encounters great obstacles because of the lack of material resources among the individual members. The need to earn a living requires them either to leave their native land in order to sell their labor-power to the highest bidder, or to remain where they are and accept the conditions, as shabby as they may be, created for them by the distribution of labor. In any case, they are enslaved, and their daily work prevents them from making plans for the future and freely choosing their allies in the battle of life. Thus it is quite remarkable that they accomplish work that may be limited in scope, but that nonetheless introduces to the world around it a new quality of life. Moreover, some signs of the society of the future do occasionally appear here and there among the workers, thanks to favorable circumstances and to the strength of the idea, which pervades even some social circles in the world of the privileged.

Often it pleases our critics to ask us sarcastically about previous attempts to create more or less communal associations in various parts of the world, and we would lack perspective if we were in any way embarrassed to answer such questions. True, the history of these associations reveals many more failures than successes. It could not be otherwise, since what is necessary is total revolution, the replacement of individual or collective work for the benefit of one, by the work of all for the benefit of all.

The persons who come together in order to form one of these societies with new ideals are themselves by no means completely rid of prejudices, old practices, and deeply rooted atavisms; they have not yet "shed the old man"! In the "anarchist" or "harmonist" microcosm they have created, they must always struggle against the dissociative and disruptive forces produced by habits, customs, the ever-powerful bonds of family, tempting advice from friends, the return of worldly ambitions, the need for adventure, and the obsession with change. Pride and a feeling of dignity can sustain novices for a while, but at the first disappointment, it is easy for them to succumb to a secret hope that the undertaking will fail and that they will once again plunge into the tumult of life on the outside. We are reminded of the experience of the colonists of Brook Farm in New England, who remained faithful to their association, if only through the bonds of virtue and loyalty to their original intention. Nevertheless, they were delighted when a fire destroyed their communal palace, thus absolving them from the agreement contracted among them in what amounted to a sort of interior vow, albeit not in the monastic sense. Obviously, the association was doomed to perish, even if the fire had not fulfilled the innermost desire of some, since the basic will of the members was at odds with the functioning of their colony.

Most communitarian associations have perished for similar reasons related to their inability to adapt to their surroundings. They were not regulated, as are barracks or monasteries, by the absolute will of religious or military masters and by the no less absolute obedience of their inferiors—soldiers, monks, or nuns. Besides, they did not yet possess the bond of complete solidarity that results in absolute respect for persons, intellectual and artistic development, and the prospect of a great and continually growing ideal. The opportunities for discord and even disunity are even more to be expected when colonists, attracted by the mirage of a distant land, are drawn toward a region completely different from their own, where each thing seems strange to them and where adaptation to the soil, the climate, and local customs is subject to the greatest uncertainties. The phalansterians who accompanied Victor Considérant to the plains of northern Texas shortly after the foundation of the Second Empire were headed toward certain ruin. They settled in the midst of populations whose brutal and coarse customs surely must have shocked their thin Parisian skins. Also, they encountered the abominable institution of black slavery and were even forbidden by law to express their opinion about it. Similarly, the experiment of Freiland, or "Land of Freedom," attempted under the direction of a Prussian officer in regions known only through vague stories and conquered with difficulty through a war of extermination, offered a farcical spectacle to the historian. It was evident from the beginning that all these heterogeneous elements would not be able to unite in a harmonious whole.

None of these failures can discourage us, for the successive efforts indicate an irresistible striving of the social will. Neither disappointments nor ridicule can deter the seekers. Besides, they always have before them the example of the "cooperatives"—consumers' associations and other types—which also had difficult beginnings but which now have become numerous and wonderfully prosperous. Undoubtedly, several of these associations have turned out very badly, especially the most prosperous among them, in the sense that the realization of profit and the desire to increase it have inflamed the love of wealth among the members of the cooperative, or at least diverted them from the revolutionary enthusiasm of the early years. This is the most formidable danger, human nature being always ready to grasp at excuses to avoid the risks of the struggle. It is easy to confine oneself to one's "good work," thrusting aside the concerns and dangers that arise from devotion to the revolutionary cause in its full scope. One tells oneself that it is especially important to succeed in an undertaking that involves the collective honor of a great number of friends, and one gradually allows oneself to be drawn into the petty practices of conventional business. The person who had resolved to change the world has changed into nothing more than a simple grocer.

Nevertheless, studious and sincere anarchists can learn a great lesson from these innumerable cooperatives that have emerged everywhere and joined to form ever larger entities in such a way as to encompass the most diverse functions, such as those of industry, transportation, agriculture, science, art, and entertainment. The scientific practice of mutual aid is spreading and becoming easy to achieve. All that remains is to give it its true meaning and morality by simplifying the whole system of exchange of services and retaining only the simple recording of statistics of production and consumption, thereby eliminating the large books of "debit" and "credit," which will have become useless.

This profound revolution is not only on the path to fulfillment, but is actually being realized in various places; however, it is pointless to draw attention to the endeavors that seem to us to be closest to our ideal, for their chances of success are greatest if silence continues to protect them, if the clamor of publicity does not disturb their modest beginnings. Let us remember the history of the small society of friends that had gathered under the name of the "Montreuil Commune." Some painters, carpenters, gardeners, housekeepers, and schoolteachers got the idea of simply working for one another, without hiring a bookkeeper as intermediary or asking the advice of a tax collector or notary. If someone needed chairs or tables, he went to see the friend who made them. If someone's house had become a bit shabby, he informed a comrade, who brought his paintbrush and bucket of paint the next day. During good weather, the members put on clean clothes, well cared for and pressed by the women citizens, and then went for a walk to gather fresh vegetables at the garden of another friend. And every day the children studied reading with the schoolteacher. It was too beautiful! Such a scandal had to stop. Fortunately, an "anarchist attack" had spread terror among the bourgeoisie, and the minister whose name recalls the "infamous laws"[8] had the great idea of offering a New Year's present to the conservatives—a decree of arrests and mass searches. The brave communitarians of Montreuil did not survive, and the most guilty—that is, the best among them—had to submit to that disguised torture called the secret investigation. And so the dreaded little commune was killed. But do not fear—it will be reborn.

13

On Vegetarianism (1901)

This essay first appeared in an earlier English translation in *The Humane Review* 1 (January 1901): 316–24, while the French version, "Le Végétarisme," was published later the same year in *La Réforme alimentaire* (March 1901): 37–45. The text was later reprinted as a pamphlet in both French and English and has been circulated up to the present time.

Highly qualified experts in hygiene and biology have done thorough research into questions relating to common foods, so I will be careful not to demonstrate my incompetence in offering my own opinion concerning animal and vegetable diet. Every man to his trade. Since I am neither a chemist nor a physician, I will make no references to nitrogen or albumin, nor will I report the results of laboratory analyses. Instead I will limit myself simply to presenting my own personal impressions, which probably coincide with those of many vegetarians. I will follow the path of my own experience, stopping occasionally for comments suggested by various small incidents.

First of all, I should say that the search for absolute truth played no role in the early impressions that turned me into a virtual or potential vegetarian when I was a small child still wearing baby clothes. I remember distinctly my horror at the sight of the shedding of blood. Once, a family member put a plate in my hand and sent me off to the village butcher, asking that I bring back some bloody scrap or other. Innocently obeying, I set out blithely to run the errand and entered the courtyard where one finds those executioners who slit animals' throats. I still remember this sinister courtyard in which terrifying men went by holding large knives that they wiped on blood-spattered smocks. On a porch hung an enormous carcass that seemed to me to take up an extraordinary amount of space.

From its white flesh a pink liquid ran in rivulets. I stood trembling and dumbfounded in this bloodstained courtyard, unable to advance and too terrified to run away. I have no idea what happened next. I have not the slightest memory. I think I was told that I fainted and that a sympathetic butcher carried me to his home. I weighed no more than one of the lambs he slaughtered each morning.

Other such incidents cast a gloom over my early years and, like my experience at the slaughterhouse, are landmarks in my young life. I can still see the sow belonging to some peasants who were amateur butchers, the cruelest kind. One of them bled the animal slowly, so that the blood fell drop by drop, for it is said that to make really good blood sausage, the victim must suffer a great deal. And indeed, she let out a continuous cry, punctuated with childlike moans and desperate, almost human pleas. It seemed as if one were listening to a child.

And in fact the domesticated pig is for a year the baby of the household, gorged with food to fatten him. He responds with sincere affection for all this care, though its only real aim is to add on a thick layer of bacon. But when there is a meeting of hearts, when the housewife charged with caring for the pig befriends her ward, pets him, pampers him and speaks to him, doesn't she appear ridiculous, as if it were absurd and almost disgraceful to love an animal who loves us! One of the strongest impressions of my childhood comes from witnessing a rural tragedy. A pig was slaughtered by a whole populace in revolt against a good old woman, my great aunt, who would not consent to the murder of her fat friend. The crowd from the village forced its way into the pigpen and then led the beast away by force to the rustic slaughterhouse where the machinery of butchery awaited. Meanwhile, the miserable lady collapsed on a stool, silently weeping. I stood next to her and watched her tears, not knowing whether I should share her grief or believe like the crowd that the slaughter of the pig was just and legitimate, dictated by both common sense and fate.

Each of us has witnessed some such barbarous act committed by the carnivore against the animals he eats, and this is especially true of those who have lived among the common people, far away from humdrum cities where everything is carefully pigeonholed or hidden away. There is no need to go to a Porcopolis[1] of North America, or into a *saladero* of La Plata[2] to gaze upon the horrifying massacres that are a precondition for the food we eat each day. But such impressions fade away in time. They give way to the baneful influence of everyday life, which tends to pull the individual in the direction of normality, while robbing him of anything that might make him into a unique being, a real person. Parents, official and informal educators, and doctors, not to mention that all-powerful person referred to as "everybody," all work together to harden the character of the child in relation to this "meat on feet," which, nevertheless, loves as we do, feels as we do, and might also progress under our influence, if it does not regress along with us.

And such regression is indeed one of the most deplorable results of our carnivorous practices, for the animals sacrificed to man's appetite have been systematically and methodically made ugly, weakened, deformed, and degraded in intelligence and moral worth. The very name of the animal into which the wild boar has been transformed has become the nastiest insult. The mass of flesh that one sees wallowing in a foul-smelling puddle is so repulsive to behold that we very carefully avoid any similarity of name between the animal and the dishes made out of it. What a difference there is between the mouflon's[3] physique and bearing as it leaps from rock to rock in the mountains, and that of the sheep, which is forever robbed of all individual initiative, reduced to mere inert flesh at the mercy of its fears. It never dares to stray from the flock and even throws itself into the jaws of the dog that pursues it. The same kind of degeneration is evident in the cow that we see moving laboriously across the pasture, transformed by its breeders into an enormous ambulatory mass, shaped geometrically as if it were designed explicitly for the butcher's knife. And it is to the creation of such monstrosities that we that we apply the term "breeding"! This is how man carries out his mission as educator in relation to his animal brethren!

Isn't this moreover the way that we act in relation to all of nature? Let loose a pack of engineers in a charming valley, in the midst of meadows and trees, or on the banks of a beautiful river, and you will soon see what they are capable of doing to it. They will do everything in their power to make their own work conspicuous and hide nature under piles of gravel and coal. They will be quite proud to see the sky crisscrossed by streaks of filthy yellowish or black smoke from their locomotives. These same engineers also sometimes claim to beautify nature. Thus when some Belgian artists recently protested against the devastation of the countryside along the Meuse River, the Minister in charge quickly assured them that he would make them happy in the future by having all the new factories built with Gothic turrets! Similarly, butchers display dismembered carcasses and bloody pieces of meat before the eyes of the public, even along the busiest streets, next to perfumed shops decked with flowers. They even have the audacity to decorate the hanging hunks of flesh with rose garlands to make them aesthetically pleasing.

When reading the news of the war in China,[4] one is amazed that the atrocities reported are a sad reality rather than a bad dream. How is it possible that men who once had the joy of their mothers' caresses and were taught in school words such as "justice" and "kindness" turn into wild beasts with human faces who take pleasure in tying Chinese people together by their clothing and pigtails and then throwing them into a river? How is possible for them to finish off the wounded and force prisoners to dig their own graves before shooting them? Who are these terrifying assassins? They are people like ourselves, who study and read as we do, who have brothers, friends, wives, and fiancées. We are likely to meet them sooner or later and shake their hands without noticing any traces of blood on them.

But isn't there a direct causal relationship between the food eaten by these executioners, who call themselves "civilizers," and their brutal deeds? They often praise bloody flesh as a source of health, strength, and intelligence. And without disgust they go into butcher shops with slippery reddish pavement and breathe the sickly sweet odor of blood! How much difference is there between the dead carcass of a cow and that of a man? Their severed limbs and entrails mixed in with one another look quite similar. The slaughter of the former facilitates the murder of the latter, especially when an order resounds from a superior, or when one hears from afar the words of his royal master, "Show no mercy!"

A French proverb says that "any evil deed can be denied." There was a certain degree of truth to this as long as the soldiers of each nation committed their acts of cruelty in isolation, for the reports of their atrocities could then be dismissed as the products of jealousy and national hatred. But today in China, the Russians, French, English, and Germans no longer conceal things discretely from one another. Eyewitnesses and even the culprits themselves inform us of them in many languages, though some do it with open cynicism and others more reluctantly. Since the truth can no longer be denied, it has become necessary to create a new morality to explain it. This morality holds there are two laws for mankind, one law for those with yellow skin and another law that is the prerogative of the whites. Apparently, in the future it will be permissible to kill or torture the former, while it will still be wrong to do so to the latter.

But isn't morality equally flexible when applied to animals? By goading dogs on to tear a fox to pieces, the gentleman learns how to send his marksmen after the fleeing Chinese. The two kinds of hunt are part of one and the same "sport," though when the victim is a man the emotion and pleasure are no doubt more intense. One might ask the opinion of the officer who recently invoked the name of Attila, pointing to that monster as a model for his soldiers!

It is in no way a digression to mention the horrors of war in connection with massacres of cattle and carnivorous banquets. People's diet corresponds closely to their morality. Blood calls for blood. On considering one's acquaintances, one usually finds that the agreeable manners, kindness of disposition, and equanimity of the vegetarians contrast markedly with the qualities of the inveterate meat-eaters and avid drinkers of blood.

Such qualities are not held in very high esteem by those "supermen" who, without actually being superior to other mortals, excel in arrogance and imagine that they advance themselves by belittling the humble and exalting the strong. In their view, the meek are weak and sickly; they block our path, and it would be a noble deed to sweep them aside. If they are not killed, they should at least be allowed to die! But it is precisely such gentle people who might well be more resistant to ills than the violent. Those with a ruddy complexion are not usually the ones who live the longest. The truly strong are not those who exhibit their strength on the surface, in a flushed face, bulging muscles, or huge, glistening bulk. Statistical studies could quickly settle this point conclusively, and would have probably done so already were there not so many biased individuals using figures, whether true or false, as ammunition to defend their pet theories.

Be that as it may, we merely contend that for the great majority of vegetarians, the question is not whether their biceps and triceps are firmer than those of carnivores, nor even whether their constitutions are better able to cope with the trials of life and the risk of death—as important as these matters may be. For them, the real concern is to recognize the bonds of affection and kindness that link man to animals. It is to extend to our brothers who have been dismissed as inferior the feelings that have already put an end to cannibalism within the human species. The arguments that cannibals once gave against the elimination of human flesh from their daily diet have the same merit as those that the typical meat-eater employs today, and the arguments that were used against that abominable custom are precisely the ones that we now present. The horse and the cow, the wild rabbit and the cat,[5] the deer and the hare—these are more valuable to us as friends than as meat. We are eager to have them either as respected fellow workers, or simply as companions in the joy of living and loving.

"Nevertheless," we are told, "if you abstain from the flesh of animals, then other carnivores, whether man or beast, will eat them instead, or else hunger and the elements will see to it that they die." Indeed, the balance of nature will be maintained as always, in accord with the hazards of life and the conflict of appetites. But at least in the struggle between species the job of destroyer will belong to others. We will develop the part of the earth that falls to us so as to make it as pleasant as possible, not only for ourselves, but also for the animals of our household. We will take seriously the role of educator that man has claimed since prehistoric times. Our share of responsibility in the transformation of the universal order does not extend beyond ourselves and our immediate environment. Even though we can do only a little, this little will at least be our own work.

If we had the chimerical idea of taking the practice of our theory to its ultimate and logical conclusion, without taking into account any other considerations, we would surely fall into complete absurdity. In this way, the principle of vegetarianism is exactly the same as any other principle; it must be adapted to the normal conditions of life. Obviously, we have no intention of dedicating all our practices and actions, every hour and every minute, to respecting the life of the infinitely small. We will not let ourselves die of hunger and thirst, like a Buddhist lama, when a microscope shows us a drop of water teeming with animalcules. We will not hesitate occasionally to cut a stick in the forest or pick a flower in a garden. We will even go so far as to use lettuce, cabbages, and asparagus for our food, although we fully acknowledge that life exists in plants as well as in animals. But we are not interested in founding a new religion and chaining ourselves to it with sectarian dogmatism. Our goal is to make our existence as beautiful as possible, and, as best we can, to adapt it to the aesthetic conditions of our environment.

Just as our ancestors became disgusted with eating their fellow humans and one fine day stopped serving them on their tables, and just as there are many carnivores today who refuse to eat the flesh of man's noble companion, the horse, or that of those pampered guests in our homes, the dog and the cat—in the same way it is repugnant to us to drink the blood and chew the muscle of the steer, an animal whose labor helps supply us with bread. We no longer want to hear the bleating of sheep, the bellowing of cows, or the grunts and piercing cries of pigs as they are led to the slaughterhouse. We look forward to the day when we will no longer have to rush quickly past hideous sites of killing to see as little as possible of the rivulets of blood, the rows of cadavers hanging from sharp hooks, and the blood-stained workers armed with gruesome knives. We hope to live one day in a city in which we no longer risk seeing butcher shops full of carcasses next to silk and jewelry stores, or across from a pharmacy, a stand with fragrant fruit, or a fine bookstore full of engravings, statuettes and works or art. We want to be surrounded by an environment that pleases the eye and is an expression of beauty.

And since we know from physiologists and even more from our own experience that this vile diet of dismembered flesh is not necessary to sustain our existence, we put aside all these awful foods that our ancestors once enjoyed and that most still do today. We hope that before long the meat-eaters will at least have the discretion to hide their food. Slaughterhouses are already banished to the outskirts of town. Butcher shops should be treated similarly. Like cowbarns, they should be tucked away in dark corners.

It is also because of their ugliness that we abhor vivisection and all dangerous experiments, except when they are heroically carried out by the scientist on his own person. And it is also because of the ugliness of the act that we are disgusted when a naturalist pins live butterflies in his specimen box or destroys an entire anthill in order to count the ants. We turn with repugnance from the engineer who defaces nature by imprisoning a waterfall in cast-iron pipes, and from the logger in California who cuts down a tree that is four thousand years old and three hundred feet high in order to show its rings at fairs and exhibitions. Ugliness in persons, in actions, in life, and in the natural environment are all the enemy par excellence. Let us become beautiful ourselves, and let our life be beautiful!

Which foods then seem most in accord with our ideal of beauty, both in their nature and in the manner in which they are produced? Precisely those that have always been most appreciated by those who lived a simple life, the foods that have no need for deceptive culinary tricks. These include eggs, grains, and fruits, in other words, the products of animal and vegetable life that represent both the temporary cessation of the organism's vitality and the concentration of the materials necessary for the formation of new life. The eggs of an animal, the seeds of a plant, and the fruit of a tree are the end of an organism that no longer exists, and the beginning of an organism that does not yet exist.[6] Man collects them for his food without killing the being that provides them since they are formed at the point of contact between two generations. Moreover, do not the scientists who study organic chemistry tell us that the egg of the animal or plant is the repository par excellence of every vital element? *Omne vivum ex ovo*.[7]

14

The History of Cities (1905)

Portions of the following appeared in an earlier English text, "The Evolution of Cities," which was published in *The Contemporary Review* 69 (January–June 1895: 246–64. Reclus' ideas concerning cities have been known primarily through that text, which was finally published in French in *Cahiers d'économie et de sociologie rurales* 8 (1992): 67–74. The present text constitutes Reclus' final formulation of his ideas concerning the city, in volume 5 of *L'Homme et la Terre* (Paris: Librairie Universelle, 1905–8), 335–76. It consists of the entire chapter on "Répartition des Hommes" ("The Distribution of Human Population").

The natural attractive force of the soil tends normally to distribute human beings rhythmically across the entire earth. In the modern period, we encounter a seemingly opposing force that concentrates hundreds of thousands or even millions of people in certain circumscribed areas surrounding markets, palaces, forums, and parliaments. Towns were already of considerable size at the outset of the age of railroads. Now, they develop into immense cities, vast agglomerations of aligned houses, crisscrossed by an infinite network of streets, alleyways, boulevards, and avenues. During the day, a grayish dome of smoke hangs over them, while at night a glow radiates outward, illuminating the sky. People were astounded by the Babylons and Ninevehs of ancient times. However, our modern Babylons, which are both cursed and celebrated, are much larger, more complex, and more teeming with humanity and gigantic machinery. Rousseau, deploring the degradation of so many country people who disappeared into the big cities, calls them "abysses" that swallow up humanity, whereas Herder sees in them "the entrenched camps of civilization." And here is how Ruskin judges them, attacking above all the largest if not the most hideous of today's cities, the capital of the immense British Empire:

> The first of all English games is making money. . . . So all that great foul city of London there,—rattling, growling, smoking, stinking,—a ghastly heap of fermenting brickwork, pouring out poison at every pore,—you fancy it is a city of work? Not a street of it! It is a great city of play; very nasty play, and very hard play, but still play. It is . . . a huge billiard table without the cloth, and with pockets as deep as the bottomless pit; but mainly a billiard table, after all.[1]

All the railing against cities by their critics is justified, as are all the encomiums of those who glorify them. How much lifeblood has gone to waste or even been destroyed by hatred, in these cities of foul air, deadly contagion, and chaotic struggle! But is it not also out of these confluences of humanity that new ideas have burst forth, new works have been born, and the revolutions that have delivered humanity from its gangrenous senility have erupted? "There is an infernal vat upon the earth," proclaims Barbier.[2] And for his part, Hugo glorifies this same Paris in enthusiastic verse: "Paris is the mother city! . . . Where generations come / To feed themselves with ideas!"[3]

The divergent tendencies of cities toward both good and evil is prefigured in the passions and will of those who flee the small towns and countryside for the big city, sometimes finding there a larger life, sometimes decline and death. But in addition to these bold forerunners who proceed resolutely toward some modern Babylon, we must count those—and they are legion—who are drawn toward centers of population and deposited there like alluvium carried by the current to be cast upon the beaches. These include peasants evicted from their plot of land for the benefit of a wealthy speculator or at the whim of a lord who decides to turn their fields into a pasture or hunting ground; servants who are summoned from the country by the city-dwellers; wet-nurses called to breast-feed infants in place of their mothers; workers, soldiers, employees, and civil servants who are transferred to the big city; and, in general, all those who in obedience to their masters, or indeed to that most imperious of masters, economic necessity, inevitably swell the urban population.

How pleasant are the words of the moralistic landowners who advise the country people to remain attached to the land, while by their actions they uproot those very peasants and create for them the living conditions that compel them to flee toward the city. Who put an end to the commons? Who reduced and then abolished entirely the rights of usufruct? Who clear-cut the forests and the moors, depriving the peasant of the fuel he needed? Who built walls around property to mark well the establishment of a landed aristocracy? And when large industry was born, did the landowner not abandon the country miller and the humble village artisans?

And when the peasant no longer has any communal lands, when he is deprived of his small industries, when all his resources are diminished at the same time that his needs and expenses grow, is his inevitable flight to the city so surprising? The landowner no longer employs full-time agricultural labor, so the worker is ruined by unemployment and forced into exile. When the proprietor needs hands for the grape harvest, he no longer looks to the old tenants of his land, but to the men of the "mobile army"—to the Irish, the Flemish, the "Gavaches,"[4] and to the anonymous workers who come from who knows where, whose birthplace, language, and customs are unknown, and who will soon disappear without leaving a trace.

Thus the immigrants drawn in multitudes toward the maelstrom of the cities obey a law that is more powerful than their own wills. Their own caprice plays only a very subordinate role in generating the force that attracts them. The relatively small number of escapees from the countryside who voluntarily head for the cities can be divided into several distinct groups. Though all may go in search of happiness, personal gain, and greater satisfaction in their emotional lives, the meaning of these ideals varies completely from individual to individual. Many of them succumb to a kind of dread that seems inexplicable. One gazes in amazement at one of their cottages, superbly situated in the mountains of the Jura, the Pyrenees, or the Cévennes. The legal owner has allowed it to fall into ruin, even though it seems to possess all the qualities that would cause one to cherish it. Alongside the dwelling rises the ancestral tree, shading the roof. Nearby, a spring of pure water gushes forth from an undulation in the meadow. Everything that can be seen from the threshold—the garden, the meadow, the fields, the groves—belonged to the family, and evidently still does. But the family now consists only of two elderly persons trying to devote their remaining energies to the farming and the household chores. In spite of this, everything perishes. The marsh encroaches on the meadow, weeds invade the paths and the flowerbeds, the harvest shrinks from year to year, and the roofs of the barns and granaries cave in. When the old people are gone, the house will collapse. But do they not have a single family member—a son, a grandson, or nephew—who might continue the work of their ancestors, as they themselves do? Yes, they have a son, but he despises the land. He has become a policeman in some distant town, taking pleasure in rounding up drunks and handing out tickets. When his parents die, he will not know what to do with the ancestral fields. They will fall fallow and a great landlord will purchase them, or rather get them for a song, to round out his hunting grounds.

If these were the only causes of the remarkable expansion of cities, they would become nothing more than social cancers, and one might justly curse them, as the Hebrew prophets once cursed ancient Babylon.

Growing by the day or even by the hour, like octopuses extending their long tentacles into the countryside, these cities indeed seem to be monsters, gigantic vampires draining the blood from men. But every phenomenon is complex. The wicked, depraved, and decadent will consume and corrupt themselves more rapidly in a milieu obsessed with pleasure or indeed fallen into decay. However, there are others with better motives, who wish to learn, who seek opportunities to think, to improve themselves, to blossom into writers or artists or even the apostles of some truth. They turn reverently toward museums, schools, and libraries, and renew their ideals through contact with others who are equally in the thrall of great things. Are they not also immigrants to the cities, and is it not thanks to them that the chariot of civilization continues to move forward through the ages? When cities grow, humanity progresses, and when they shrink, the social body is threatened with regression into barbarism.

Without having studied the question, one might easily imagine that cities are distributed randomly. And in fact, a number of accounts depict the founders of cities leaving to fate the choice of a site on which to settle and build protective walls. The course of the flight of birds, the spot on which a stag was hunted down and taken, or the point at which a ship ran aground determined where a city was to be constructed. Thus the capital of Iceland, Reykjavik, is supposed to have been founded according to the will of the gods.[5] In 874, the fugitive Ingolfur came in sight of Iceland and cast into the water the wooden images that served as his household idols. He sought vainly to follow their course, but they eluded him, and he had to establish a temporary camp on the shore. Three years later, he rediscovered the sacred pieces of wood, and moved his settlement to a nearby site, which turned out to be as favorably situated as possible in this formidable "Land of Ice."[6]

If the earth were completely uniform in relief, in the quality of its soil, and in its climatic conditions, cities would be distributed in geometrical positions, so to speak. Mutual attraction, social instinct, and convenience for trade would have given rise to them at equal distances from one another. Given a region that is flat, that has no natural obstacles, rivers, or ports, that is situated in a particularly favorable manner, and that is not divided into separate political states, the largest city would be constructed precisely at the center of the country. The secondary cities would be distributed at equal intervals around it, spaced rhythmically. Each of these would have its own planetary system of smaller towns, and each of these its retinue of villages. On a uniform plain, the interval between the various urban agglomerations should be the normal distance of a day's walk. The number of leagues that could be covered by the average walker between dawn and dusk—that is to say, between twelve and fifteen, corresponding to the hours of the day—constitutes the usual distance between towns. The domestication of animals, then the invention of the wheel and finally machines have modified, either gradually or abruptly, these early measurements. The gait of the horse, and later the turn of the axle determined the normal distance between the great gathering places of humanity. The average interval between villages is measured by the distance covered by a farmer pushing his wheelbarrow full of hay or grain. A supply of water for the cattle, the convenient transportation of the fruits of the earth—such factors determine the site of the stable, the granary, and the cottage.

In a number of countries that have been populated for a long time, and in which the distribution of the urban population is still in accord with the original distances, one finds beneath the apparent disorder of cities an underlying order of distribution that was clearly determined long ago by the footsteps of walkers. In the "Middle Flower,"[7] in Russia, where railways are a relatively recent creation, and even in France, one can observe the astounding regularity with which the urban agglomerations were distributed before mining and industry came to disturb the natural equilibrium of population.[8] Thus Paris, the capital of France, is surrounded, in the direction of the country's borders and coastlines, by cities that are second only to itself in importance: Bordeaux, Nantes, Rouen, Lille, Nancy, and Lyons. The ancient Phoenician and later Greek city of Marseilles owes its origins to a different phase of history from that of the cities that were Gallic and then French. Nevertheless, its position corresponds to theirs since it is situated at the Mediterranean extremity of a radius that is twice the distance between Paris and the great urban planets in its orbit. Between the capital and the secondary administrative centers, cities of considerable though lesser importance, such as Orléans, Tours, Poitiers, and Angoulême, were founded. These were established at approximately equal intervals, for they are separated by a double daily traveling distance, that is, between twenty-five and thirty leagues. Finally, halfway between these tertiary centers, modest towns such as Etampes, Amboise, Châtellerault, Ruffec, and Libourne took shape. Their locations marked an average day's traveling distance. Thus the traveler crossing France found alternately a town that was a simple resting place and a town with all the amenities. The first was adequate for the traveler on foot, while the second was suitable for the rider. On almost all the highways, the rhythmic distribution of cities occurs in the same manner, through a natural cadence regulated by the pace of men, horses, and carriages.

The irregularities in the network of settlements are all explained by such factors as the contour of the land, the course of rivers, and the thousand variations of geography. In the first place, the nature of the soil determines where people choose sites for their dwellings. The village can only spring up where the stalk sprouts. People turn away from barren heaths, masses of gravel, and heavy clays that are difficult to plough, and rush immediately and spontaneously to areas of loose soil that is easy to work. They also avoid low, moist regions, although these have an exceptional fertility. The history of agriculture shows that these soft alluviums repel people because of their unhealthiness. They have been cultivated through collective efforts that only become possible when humanity has advanced considerably.

Terrain that is too uneven and soil that is too arid also fail to attract population, thereby preventing or delaying the establishment of cities. Glaciers, snow, and cold winds expel people, so to speak, from the harsh mountain valleys. The natural tendency is to found cities immediately outside such forbidding regions, at the first favorable spot available—for example, just at the entrance to a valley. Every stream has its riparian city in the lowlands, where the riverbed suddenly widens and divides into a multitude of branches through the gravels. Every double, triple, or quadruple confluence of valleys gives rise to a large agglomeration whose size is proportional, all things being equal, to the volume of water carried by the convergent riverbeds. Could any site for a city be more naturally determined than that of Saragossa, which is in the middle of the course of the Ebro, at the junction of the double valley through which the Gállego and the Huerva flow? Similarly, the city of Toulouse, the metropolis of the Midi of France, occupies a site that even a child could have pointed out as a likely meeting place for peoples, just where the river becomes navigable, below the confluence of the upper Garonne, the Ariège, and the Hers. At the two western corners of Switzerland, Basel and Geneva were built at the crossroads of the great paths followed by migrating peoples. And on the southern slope of the Alps, every valley without exception has at its entrance a guardian town. Powerful cities such as Milan and so many others mark points of geographical convergence. The upper valley of the Po, constituting three-quarters of an immense circle, has at its natural center the city of Turin.

On the lower course of the river the establishment of cities is determined by conditions analogous to those that prevail at the middle. It occurs at the headland of two streams, at the ramification of three or four navigable waterways or natural routes that come together, or at the point on a river where it intersects with natural land routes leading in various other directions. In addition, other groups settle at necessary stopping places, such as rapids, waterfalls, or rocky gorges, where boats drop anchor and where merchandise is transshipped. The straits of rivers and any spots where the crossing from bank to bank is particularly easy are also appropriate for the site of a village or even a town, if there are additional advantages besides the narrowing of the river. If a marked bend in a waterway brings its valley into close proximity to a large center of activity situated in another basin, this can also attract a large number of settlers. Accordingly, Orléans had to be built on the bank of the Loire conducive to expansion toward the north in the direction of Paris, and Tsaritsin[9] is located at the place where the Volga is closest to the Don. Finally, on every river the vital point par excellence is the area around its mouth, where the rising sea stops and supports the upper current and where the smaller boats, carried by the current of fresh water, naturally meet the seagoing vessels coming in with the tide. In the hydrographic organization, this meeting place can be compared to the collar of a tree between the aerial vegetation and the underground root system. This is the normal pattern for the large European tidal seaports such as Hamburg, London, Antwerp, and Bordeaux.

The irregularities of the coastline also affect the distribution of cities. Certain sandy shorelines with little variation, inaccessible to ships except on those rare days of complete calm, are avoided by people from inland as well as by those who sail the seas. Thus the 220-kilometer coastline that runs in a straight line from the estuary of the Gironde to the mouth of the Adour has not a single town other than little Arcachon, which is no more than a simple bathing spot and resort, situated away from the shore within a rampart formed by the dunes of Cape Ferret. Similarly, the impressive barrier islands that follow the Atlantic coast of the Carolinas allow access between Norfolk and Wilmington only to a few poor towns that carry on a hazardous trade with considerable difficulty. In other coastal regions, islands and islets, rocks, promontories, and peninsulas multiply the thousand jagged edges and gashes of the escarpments. These similarly prevent the birth of towns, despite the advantages offered by deep and well-protected waters. Where coastlines are too violent and tempestuous, only a small number of people will be able to settle easily. The most favorable sites are those that have a temperate climate and are accessible from both land and sea, by ships and vehicles of all kinds.

In contrast to the regular coast of the Landes, which is almost devoid of towns and villages, one can point to the Mediterranean coastline of Languedoc between the delta of the Rhone and the mouth of the Aude. In this region the large centers of population are found in closer proximity than they are on average anywhere else in France, even though the density of population per square kilometer is no greater than that of the country as a whole. The explanation for this string of cities is to be found in the geographical features of the countryside. The route that those traveling from Italy used to follow to reach Spain or Aquitaine had to avoid both the steep mountains of the interior and the marshes, salt lakes, and mouths of rivers along the coast. The steep, sparsely populated, and rather inhospitable upland area that borders the mountainous wall of the Cévennes to the south begins in the vicinity of the sea. Historically, movement through the region has thus shifted to a route that follows the Mediterranean coast. On the other hand, trade requires points of access, whether they be the mouth of a river such as the Aude or the Hérault, or else a cove artificially protected by jetties. Such considerations are responsible for the establishment of Narbonne, which enjoyed a period of world power when it was the most populous city of Gaul; Béziers, which prospered during the Phoenician period and which remains one of the great agricultural markets of France; Agde, the Greek town, which was succeeded in importance by Sête, another town with Hellenic origins; and Montpellier, the intellectual capital of the Midi, where the Saracens and the Jews were the precursors of the Renaissance. Beyond, other towns crowd together. The ancient Nîmes, sitting beside its fountain, is linked with the Rhone through the three cities of Avignon, Beaucaire, and Arles.

All natural conditions, including agricultural, geographical, and climatic ones, influence the development of cities, whether for better or for worse. Every natural advantage increases their powers of attraction, and every disadvantage diminishes them. Given the exact same historical environments, the size of cities would be directly proportional to the sum of their natural endowments. However, two cities, one in Africa and the other in Europe, might have similar natural environments yet differ considerably from one another because the context of their historical evolution is so different. Nevertheless, there will be similarities in their destinies. And just as celestial bodies affect one another, neighboring urban centers mutually influence one another. They may either work together because they offer complementary advantages, as is the case with the commercial city of Liverpool and the manufacturing city of Manchester, or harm one another when they each have the same benefits to offer. The latter is the case with Bordeaux, on the Garonne, and Libourne, on the Dordogne, which are situated not far apart, on the two sides of the "Entre-deux-Mers." Libourne could have offered almost identical services to trade; however, its proximity to Bordeaux hurt its chances. Devoured by its rival, it lost virtually all its maritime significance and has no importance today other than as a stopping place for travelers.

Another remarkable phenomenon that should be noted is the ability of geographical forces, much like those of heat and electricity, to act at a distance, producing effects far from their source. Thus a city may rise up on a certain site as the result of various factors that make it preferable to sites closer to that source. One can cite the examples of three Mediterranean ports located where river deltas have created conditions that are particularly appropriate for trading cities. Despite its distance from the Nile, Alexandria serves as the commercial center for the entire river basin, while Venice is the port for the Paduan plain, and Marseilles, for the valley of the Rhone. And though Odessa is twenty kilometers from the mouth of the Dnieper, it still oversees the river's trade.

In addition to the qualities of the climate and the soil, those of the subsoil sometimes exercise a decisive influence. A city may rise up suddenly at a seemingly inhospitable spot, thanks to the area's subterranean wealth in building stones, clay for molding and sculpting, chemicals, various metals, and combustible minerals. Thus Potosí, Cerro de Pasco, and Virginia City have sprung up in regions that, apart from the presence of silver deposits, could never have supported a city. Merthyr Tidfil, Le Creusot, Essen, Liège, and Scranton are creations of coal mines. Formerly unused forces of nature are now giving birth to new cities in places that were once shunned, such as at the foot of a waterfall, as in the case of Ottawa, or in mountainous areas that are now within reach of electrical lines, as in the valleys of Switzerland. Every advancement by man creates points of vitality in unexpected places, much as a new organ creates its own nerve centers. What rapid changes in the distribution of cities are in store when man will have become the master of aeronautics and aviation! Just as man now seeks new sites along seacoasts that are most capable of handling the coming and going of ships, in the future he will feel as if he were carried like an eagle toward the summits from which his view can embrace the infinity of space.

To the degree that the sphere of human consciousness expands and interactions occur across much greater distances, cities become members of a greater organism. To the particular advantages that caused their birth, they add assets of a more general nature that may allow them to play a major role in history. Thus Rome, Paris, and Berlin have never ceased to gain new causes of growth, including growth itself.[10] Can we not say this of London, today the largest city in the world? The principal cause of its prosperity is its location as a port, being situated at the head of maritime navigation of the Thames. This has allowed the city, which became the capital of the United Kingdom, to develop various assets that might otherwise have remained mere possibilities, never to be realized. Thus, advancing further and further in relation to the rest of the world, London has ended up becoming the central point that is on the whole most easily accessible from every corner of the earth.

As cities develop, it often happens that the growth or decline of these great organisms moves irregularly, by fits and starts caused by rapid historical change. To take the example of London again, one can see that at the outset, the local advantages of the city, while having a certain importance, could not in themselves explain the rank that it has achieved among the world's cities. Many conditions were most favorable to London in helping it prevail in its struggle with other English cities for survival. It is located on a plain that is clearly bordered on the north by protective hills. It is on the banks of a great river and at the confluence of another smaller waterway. And it is positioned at the very point where the rise and fall of tides facilitates the alternation of navigational direction and the loading and unloading of merchandise. However, these local advantages would never have realized their true value had the Romans not chosen this site as the central convergence of the routes extending in every direction across the southern half of the great island. The British Rome could only rise up on the site chosen as the center of this network. But when the Roman legions had to abandon Albion and all the "high streets" constructed between the military posts and the country's port were deserted, Londinium thereby lost all of its importance. It became no more than a simple British village, reduced, like so many others, to dependence on its purely local assets, and for two hundred years it was completely ignored by history.[11] In order for the city to regain its significance, it was necessary that it reestablish its relationship to the continent.

The development of capital cities is to a large extent artificial. Administrative favors, the demands of courtiers and courtesans, civil servants, police, soldiers, and the self-interested multitude that crowds around the "ten thousand at the top," give capitals certain peculiar qualities that that prevent them from being studied as typical urban centers. It is easier to comprehend the life of those cities whose histories depend almost entirely on their geographical environment. No study is more fruitful than the biography of a city whose appearance, even more than its historical records, allows one to observe the successive changes that have unfolded from century to century, following a certain rhythm.

In the mind's eye one can visualize the huts of the fisherman and gardener beside one another. Two or three farms are scattered across the landscape and a millwheel turns under the weight of the tumbling water. Later, a watchtower rises upon the hill. On the other side of the river, the prow of a ferry touches the shore, and another hut is built. Beside the boatman's cottage, an inn and a shop beckon travelers and passers-by. Then a market rises up on the leveled terrace nearby. A widening path, which is increasingly beaten by the footsteps of men and animals, descends from the plain to the river, while a winding trail cuts through the hillside. Future roads begin to take shape in the trodden grass of the fields, and houses occupy the four corners of the crossroads. The chapel becomes a church, the watchtower a fortified castle, a barracks, or a palace. The village grows into a town and then a city.

The correct way to study an urban agglomeration that has gone through a long period of historical development is to examine it in detail, paying careful attention to the conditions of its growth. One should begin with the place that was its cradle, a site almost always consecrated by legend, and end with today's factories and garbage dumps.

Each city has its unique individuality, its own life, its own countenance, tragic and sorrowful in some cases, joyful and lively in others. Successive generations have left each with its distinctive character. And each constitutes a collective personality whose impression on each separate person may be good or bad, hostile or benevolent. But the city is also a very complex individual, and each of its various neighborhoods is distinguished from the others by its own particular nature. The systematic study of cities, which examines both their historical development and the social values expressed in their public and private architecture, allows one to judge them as one judges individuals. One can note the dominant elements in a city's character and judge the extent to which its influence has on the whole been either useful or detrimental to the progress of the populace that lives within its sphere of activity. Many cities are quite obviously devoted to work, but some of these differ markedly from others, according to whether local businesses operate in a normal or a pathological manner: whether they develop in conditions of peace, relative equality, and mutual tolerance, or whether they are instead carried away by the turmoil of furious competition, chaotic speculation, and brutal exploitation of the working class. Some cities can be seen immediately to be banal, bourgeois, routine, lacking in originality, and lifeless. Others are clearly designed for domination and overwhelm the surrounding countryside. They are tools of conquest and oppression, and on seeing them one experiences feelings of spontaneous horror and dread. Other cities seem completely antiquated even in their modern sections. They are places of shadow, mystery, and fear, where one feels overcome by feelings of another age. On the other hand, some cities seem eternally young. They inspire joy, their humblest structure has originality, the homes are cheerful, and the inhabitants have a poetic air and contribute to humanity their own, unique way of life. Finally, there are all the cities that have many faces, in which each social class is found in distinct neighborhoods that reflect its condition, and where attitudes and language change only slowly over the centuries. There are so many unhappy places that would make one weep!

The differences between cities are exhibited clearly in their respective modes of growth. Cities extend their suburbs outward along the highways, like tentacles that reach out in the direction of the greatest land commerce. Similarly, if a city runs along a river, its growth extends along the banks, where the boats anchor and unload. There is sometimes a striking contrast between two neighborhoods along a river that seem equally suited for human habitation, but which differ markedly because of the direction of the river's current. Thus, considering the city of Bordeaux spatially, one would conclude immediately that the real center of population should be on the right bank of the river, at a spot where the houses of the small suburb of la Bastide rise up. But here there is a large bend in the Garonne, and consequently the docks are all located along the left bank, following the more rapid current of the river. The side on which the river truly flows also carries the current of commercial and political activity. The population follows the course of the river and avoids the muddy shores of the right bank. Big business did the rest by taking over the suburb, hemming it in with intersecting circles of railroad tracks and crossing gates and defacing it with sheds and warehouses.

It has often been contended that cities have a tendency to grow westward. This phenomenon, of which there are many cases, can be explained very well in the countries of Western Europe and in those with a similar climate. In these countries the prevailing winds blow from the west. The inhabitants of neighborhoods receiving fresh air are less exposed to health hazards than those living on the other side of the city, where the air is polluted when passing over chimneys, sewers, and many thousands or even millions of human beings. Furthermore, one must remember that the rich, the idle, and the artistic who enjoy contemplating the beauties of the heavens have more occasion to do so at dusk than at dawn. They unconsciously follow the direction of the sun in its westerly movement, and take pleasure in the evening at watching it set among the radiant clouds. Yet how many exceptions there are to this normal tendency of cities to grow in the direction of the sun's path! The form and contour of the land, the appeal of beautiful sites, the direction of the currents of waterways, and the growth of neighborhoods parasitical on the needs of industry and commerce often draw people of wealth and leisure to parts of the city other than those that lie to the west. Brussels and Marseilles are two examples of such divergence from the normal model.

By the very fact of its growth and development, the urban agglomeration tends to die, like every organism. It is subject to the ravages of time, and one day discovers that it is old, while other cities are rising up, eager to live their own lives. Doubtless, because of the forces of inertia and routine among its inhabitants, and the powerful attraction that a center exerts over surrounding areas, it still maintains certain enduring qualities. But not only is the urban organism subject to the fatal accidents that befall cities as well as men, it is unable to rejuvenate and recreate itself quickly except by means of ever-greater efforts—and even then it may shrink from this continual necessity. The city must enlarge its streets and squares, rebuild, move or raze its walls, and replace old, outmoded structures with new ones adapted to changing needs.

Whereas a new American city is born fully adapted to its environment, a city like Paris, which is old, congested, and polluted, must constantly reconstruct itself. Because of this continuous effort, the city is at a great disadvantage in the struggle for existence, as compared to new cities such as New York and Chicago. It is for similar reasons that in the basins of the Euphrates and the Nile immense cities like Babylon, Nineveh, and Cairo have successively relocated. Thanks to the advantages of its site, each of these cities has retained its historical importance, at least to some degree. However, they all found it necessary to abandon certain antiquated quarters and move further on in order to avoid the debris, not to mention the stench emanating from garbage piles. In general, the only inhabitants of the site that was forsaken when the city moved on are those in the graves.

Other causes of the death of cities, more decisive because they arise from historical development itself, have struck many formerly famous cities. Conditions similar to those that gave birth to the city have been the cause of its inevitable destruction. Thus the replacement of one highway or crossroads by other roads that are more convenient can result in the elimination of a city that owed its existence to transportation. Alexandria ruined Pelusium.[12] Cartagena in the West Indies returned Portobello to the solitude of the forest.[13] The requirements of commerce and the suppression of piracy have changed the location of many cities built on the rocky coast of the Mediterranean. Once they were perched on rugged hills and encircled by thick walls to defend them from warlords and privateers. Now they have come down from their rocky heights and extend along the seashore. Everywhere the *borgo* has become a *marina*. The Piraeus[14] has succeeded the Acropolis.

In our authoritarian societies, in which political institutions have often given preponderant influence to a single will, it has sometimes happened that the whims of a sovereign have placed cities in areas in which they would never have grown up spontaneously. Once established in such unnatural environments, they have only been able to develop at the cost of an enormous loss of vital energy. Thus cities such as Madrid and St. Petersburg were built at great expense, though the original huts and hamlets, left to themselves without the actions of Charles the Fifth and Peter the Great, would never have become the populous cities that they are today. Although these cities were created by despotism, because of men's collective labor they are nevertheless able to live as if they had a normal origin. Though the natural features of the landscape did not destine them to be centers of population, they have become so because of the convergence of highways, canals, railways, transportation links, and intellectual exchanges. Geography is not an unchanging thing, but rather something that makes and remakes itself constantly. It is continually modified by the actions of men.

Today it is no longer such Caesars who build capitals; they have been succeeded by powerful capitalists, speculators, and presidents of financial syndicates. Construction covering wide areas rises up in just a few months, laid out beautifully and provided with excellent facilities; even the schools, libraries, and museums lack nothing. If the choice of sites is wise, these new creations quickly enter the mainstream of life. Thus Le Creusot, Crewe, Barrow on Furness, Denver, and La Plata have taken their place as centers of population. But if the site is poorly chosen, then the city dies along with the special interests that gave birth to it. Cheyenne, no longer the final stop on the railway, sends its little houses further down the line, and Carson City disappears when the silver mines that attracted people to the forbidding desert around it are exhausted.[15]

Not only do the whims of capital sometimes give rise to cities that are doomed by the general interests of society; they also destroy many communities whose inhabitants would be quite content to continue to live there. Do we not see, on the outskirts of many large cities, rich bankers and landowners increasing their domain each year by hundreds of hectares, systematically changing cultivated land into plantations or parks for pheasants or large game? They level whole hamlets and villages to replace them with widely scattered caretakers' huts.

One should mention, among the cities that are partially or entirely artificial and do not fulfill the real needs of industrial societies, those cities created for war, or at least those built in recent times by large centralized states. This was not the case when the city included the entire tribe or constituted the natural core of the nation. It was then absolutely necessary for defense to build ramparts that followed exactly the exterior outline of the neighborhoods, and to build watchtowers at the corners. In this period, the citadel, where all the citizens took refuge in times of grave danger, also served as the temple, and was built at the summit of the guardian hill, a monument made sacred with statues of the gods. In the case of cities like Athens, Megara, and Corinth, which consisted of two separated sections, it was necessary to protect the connecting road with long parallel walls. The arrangement of the fortifications was determined by the nature of the landscape and blended in a harmonious and picturesque manner with the countryside.

But in our day of extreme division of labor, in which military forces have become practically independent of the nation and no civilian would dare to interfere in questions of strategy, most of the fortified cities have extremely ugly contours. They have not the slightest attunement to the undulations of the landscape but instead cut up the landscape along lines that are offensive to the eye. The Italian engineers of the Renaissance, and later Vauban and his emulators, at least tried to design the outline of their fortified sites with the goal of perfect symmetry. Some of their works take the form of a starred cross with rays and gems. The white walls of their bastions and redans[16] contrast consistently with the calm quietude of the shady countryside. But our modern sites no longer aspire to beauty. This goal never enters the minds of the builders. Indeed, a mere glance at the map of a fortified town shows it to be ugly, hideous, and in complete conflict with their environment. Rather than embracing the contours of the land and freely extending its arms into the countryside, it seems as if its limbs are amputated and its vital organs stricken. Just look at the sad outward appearance of cities such as Strasbourg, Metz, and Lille! The latter is so narrowly confined within its ramparts that it had to overflow, so to speak, these military restraints. Roubaix and Tourcoing adjoin the fortified center, and today an attempt is made to merge the three elements into a harmonious whole by means of wide boulevards. Despite its beautiful buildings, its graceful promenades, and the charm of its people, Paris is another city that is marred by a harsh ring of fortifications. If the city had been freed from this unpleasant oval of broken lines, it would have grown organically, in an aesthetically pleasing and rational manner. It would have followed the more elegant contours given to it by life itself.

Another cause of ugliness in our modern cities is the invasion of large manufacturing industries. Almost every urban agglomeration is darkened with one or two areas that bristle with stinking smokestacks and are crisscrossed by gloomy streets lined with hulking structures whose walls are either completely blank or are riddled with countless depressingly uniform windows. The ground shakes under the weight of trucks and freight trains and from the effects of machinery in motion. There are so many cities, especially in young America, where the air is almost unbreathable, where everything that one encounters—the soil, the roads, the walls, the sky—seems to exude mud and soot! One can only recall with horror and disgust a mining community like the endlessly winding Scranton, whose seventy thousand inhabitants lack even a single hectare of filthy grass or sooty foliage to soothe their eyes after all the hideousness of the factory. Or consider the enormous Pittsburgh, with its semi-circular crown of elevated districts that flame and fume. Although the natives claim that the streets have become cleaner and the view clearer since the introduction of natural gas in the factories, can one imagine a filthier atmosphere? Other less blackened cities are still almost as hideous because of the railroads, which have taken over streets, squares, and walkways, and send locomotives snorting and hissing by, scattering the crowds in their path. In fact, some of the most beautiful sites on earth have been desecrated. Thus in Buffalo people try in vain to walk along the banks of the superb Niagara River, running into foundries, railway crossings, muddy canals, piles of gravel and garbage, and all the other refuse of the city.

Barbarous speculation has also ruined the streets by creating subdivisions on which contractors build large districts, planned beforehand by architects who have never even visited the site, much less gone to the trouble of consulting the future inhabitants. They erect a Gothic church for the Episcopalians, a Romanesque structure for the Presbyterians, and finally a sort of Pantheon for the Baptists. They lay out the streets in squares and diamonds, varying bizarrely the geometrical design of the public squares and the style of the houses, while religiously saving the most valuable corners for the most unsavory drinking establishments. These are contrived cities that are based on the most banal concepts and that always betray in some manner the ostentatious arrogance of their creators.

In any case, every new city immediately constitutes, by its configuration of dwellings, a collective organism. Each cell seeks to develop in perfect health, as is necessary for the health of the whole. History demonstrates that sickness is no respecter of persons; the palace is in danger when the plague rages through the slums. No municipality can ignore the importance of the thorough rehabilitation of the city through street cleaning; the establishment of parks with lawns, flowers, and large shade trees; the rapid disposal of all refuse; and the supply of an abundance of pure water to every house in every neighborhood. In this regard, the cities of the most advanced countries are in friendly competition to test and put into practice various procedures to improve cleanliness and convenience. It is true that cities, like states, have rulers whose milieu induces them to place their own self-interest above everything else. We have nevertheless achieved a great deal if we know what can be done so that some day the urban organism will function automatically to provide food, pure water, heat, light, energy, and ideas; to distribute equipment; and to dispose of useless or harmful materials. This ideal is still far from being realized. Still, many cities have already become healthy enough so that the average quality of life exceeds that of many rural areas in which the inhabitants constantly breathe the odor of rot and manure, and remain in primitive ignorance of basic hygiene.

The level of consciousness present in urban life is also expressed in a concern for art. Like Athens in ancient times, and like Florence, Nuremberg, and the other free cities of the Middle Ages, every modern city seeks to beautify itself. Even the most humble village has a bell tower, a column, or a sculptured fountain. But how sad and dreary is this art in general, concocted by highly certified professors under the supervision of a committee of incompetents whose pretentiousness is directly proportional to their ignorance. True art is always spontaneous and can never adapt itself to the dictates of a public works commission. These small-minded city council members often proceed in the style of the Roman General Mummius, who enthusiastically commanded his soldiers to repaint every damaged picture. They imagine that symmetry will achieve beauty, and think that identical reproductions will give their towns a Parthenon or a St. Mark's. In Europe we have a city whose very buildings render it preeminently banal—namely, the vast city of Munich, which contains many scrupulous imitations of Greek and Byzantine monuments, masterpieces that lack their appropriate environment, atmosphere, soil, and people.

Even if the imitators were able to produce monuments that were exact copies of their models, their works would be no less contrary to nature. A building can be understood only in relation to the conditions of time and space that gave rise to it. Each city has its own life, its particular qualities, its distinct countenance. With what great reverence architects should look upon it! It is an assault on the collective personality embodied in the city to destroy its individuality in order to litter it with unimaginative structures and monuments that clash with its present character and its past history! The true art is to adapt the contemporary city to the demands of modern labor while preserving all the picturesque, unique, and beautiful qualities it has inherited from past centuries. We must learn how to sustain the life of the city and endow it with perfect health and utility, in the same way that loving hands restore the well-being of a sick person.

Thus in Edinburgh, intelligent men who are at once artists and scientists have undertaken the restoration of the splendid thoroughfare called High Street, which extends from Edinburgh Castle to Holyrood Palace, joining the two main sections of the old city. On the departure of King James for England, it was abandoned immediately by all the parasites of the court: chamberlains, soldiers, pleasure-seekers, purveyors, and lawyers. This avenue of sumptuous mansions then had new residents, for the poor moved in, doing their best to adapt the huge rooms by dividing them up with crude partitions. Two hundred years after the desertion of the street, it had become a collection of hovels with foul-smelling courtyards and tiny rooms infested with fever. The populace, clothed in filthy rags and constantly covered with mud, consisted in large part of the infirm, the scrofulous, and the anemic. The elegant vices of the court were succeeded by the most repugnant public ones. It is these awful cesspools that the renovators have attacked, gradually transforming each house, reinstalling the wide staircases, restoring the large rooms with monumental fireplaces, bringing an abundance of fresh air and light everywhere, supplying plenty of water to even the humblest attic, and adding bas-reliefs and decorative details to the bare walls of buildings. The picturesque qualities of old structures are respectfully preserved, and are even accentuated by means of towers, pinnacles, and belvederes, while the horrible filth and stench are removed. The street that was formerly bedecked with tattered rags now contains balconies decorated with flowers and foliage. The city reemerges fresh and new, just as in a garden a trampled flower springs back with the stem and soil undisturbed.

But in a society in which people cannot depend on having enough bread to eat, in which the poor and even the starving make up a large part of the population of every large city, it is no more than a halfway measure to transform unhealthy neighborhoods if the unfortunate people who previously inhabited them find themselves thrown out of their former hovels only to go in search of new ones in the suburbs, merely moving the poisonous emanations a certain distance away. Even if the council members of a city were without exception men of impeccable taste and every restoration or rebuilding were carried out in a manner that is beyond reproach, there would still exist everywhere the painful and disastrous contrast between wealth and poverty, which is the inevitable result of inequality, the antagonism that cuts the social body in half. The counterpart to the arrogantly imposing neighborhoods is the sordid dwellings that, behind their low and leaning outer walls, conceal slimy courtyards and unsightly piles of stones and scraps of wood. Even in cities in which the administrators try to veil all these horrors hypocritically by hiding them behind decent whitewashed fences, the misery breaks through nonetheless. Behind them, death carries out its work even more cruelly than elsewhere. Is there among our modern cities a single one that does not have its Whitechapel or its Mile End Road?[17] As beautiful and imposing as an urban agglomeration may be in its entirety, it always has its open or hidden vices, its defects, and its chronic sicknesses. These will lead inevitably to death if healthy blood does not once again freely circulate throughout the organism.

How very far are so many of today's cities from such a future state of well-being and beauty. A chart published in the city directory of St. Petersburg for 1892 gives a striking example of the manner in which such a large capital city can consume human lives. Starting with the year 1754, when the population was about 150,000, over the next 126 years the rate of growth increased to the point that there were 950,000 inhabitants. However, the hypothetical rate of change, calculated according to mortality and not taking into account immigration, results in a loss of 50,000. Births do not outnumber deaths even slightly until 1885, a year of extensive sanitation projects. And across the world, how many cities, like Budapest, Lima, and Rio de Janeiro, would be on the road to quick destruction were it not for the people from the country who come to take the place of those who die! If Parisian families die out after two or three generations, is it not the pernicious odor of the city that gets to them? If young Polish Jews fail the military physicals in much greater numbers than young people of other nationalities, should the blame not be placed on the cities that condemn them to stagnate in poverty-stricken ghettos?

And in how many cities does the sky seem to be draped with a funeral veil! On entering a hazy city such as Manchester, Seraing, Essen, Le Creusot, or Pittsburgh, one can see clearly how the works of Lilliputian humans are capable of tarnishing the sunlight and profaning the beauty of nature. If a very minute quantity of coal dust escapes from combustion and produces a continuous layer of haze a fraction of a millimeter in thickness, this suffices, especially if there is fog, to counteract the light of the sun.[18] The impenetrable atmosphere that sometimes weighs on the city of London is justly famous.

Moreover, the cleaning-up of urban centers gives rise to a number of other problems, apart from that of smoke, that should be on the whole easy to solve. Unfortunately, we are far from having found effective and standardized methods for the disposal of sewage and household garbage, and for the purification of sewage water, either by chemical treatment or by its rational use in agriculture, and too many municipalities seem not even to be concerned with such questions. The adoption of road surfaces that produce neither dust nor mud, and, in general, the efficient organization of transportation, also have an important influence on public health.

Many indicators show that the flow of rural population toward the cities could come to a halt or even reverse direction. First of all, the high rent in urban areas naturally causes workers to move to the outer suburbs, and the bosses of industry can only encourage this exodus, since it will lead to a decrease in the cost of labor. The bicycle, the morning trams, and commuter trains have allowed many thousands of factory and office workers to find more affordable housing in an atmosphere that is less polluted with carbonic acid. Thus in Belgium the rural communes in many districts have maintained their population, thanks to the use of "weekly coupons." In 1900 there were no less than 150,000 workers who spent nights and Sundays in their villages, but traveled even fifty kilometers, at a weekly cost of two francs twenty-five centimes, to work every weekday in a workshop or factory in some distant city. But this is a spurious solution since the head of the family exhausts himself through long journeys, bad meals, and shortened nights of rest, and besides, the villages have the same health and sanitation problems as the cities.[19]

And this is not all. The electricity generated by waterpower tends to replace coal as an energy source, so that factories are scattered along waterways. Thus Lyons, despite the strong attraction of its industry and arts, nevertheless shrinks by several thousand inhabitants each year. This is not because it is becoming less prosperous, but on the contrary, because its rich textile manufacturers and other industrialists have extended their sphere of activity to all the surrounding *départements*, and even as far as the Alps—anywhere that waterfalls and rapids offer them the energy resources they require.

To judge things correctly, we must recognize that every question of municipal governance is inseparable from the social question itself. Will we see the day when all people without exception can breathe fresh air, enjoy the full sunlight, delight in the pleasant shade, savor the fragrance of roses, and generously provide for their families without fearing that they cannot put food on the table? When this day comes, and only then, cities will be able to realize their ideal and transform themselves in a manner that corresponds exactly to the needs and desires of all. They will finally become perfectly healthy and beautiful organic bodies.

This is the avowed goal of the Garden City.[20] Indeed, intelligent industrialists and innovative architects have succeeded in creating in England, where urban blight has been the most hideous, a certain number of centers in which conditions are equally healthy for all, including the poor as much as the rich. Port Sunlight, Bourneville, and Letchworth certainly offer a pleasant alternative to the slums of Liverpool, Manchester, and similar cities. The low mortality rates for these new towns rival those of the most opulent neighborhoods of our great capitals—only ten to twelve deaths annually per thousand inhabitants. But it is still the privileged who live in the Garden Cities, and the good will of all the philanthropists in the world is not sufficient to conjure away the antagonism that exists between Capital and Labor.

Long before these experiments of our own day, we find in many villages of our ancestors touching evidence of the quest for a beauty that could only be satisfied by the creation of a harmonious whole. One can cite notably the communities of the Polabians,[21] a people of Slavic origin who live in the valley of the Jeetze, a branch of the Elbe in Hanover. All the houses are spaced around a central oval plaza containing a small pond, a grove of oaks and lime trees, and some stone tables and benches. Each dwelling is dominated by a high gable supported by a projecting framework. Its facade is turned toward the plaza, and above the door there is an inscription with moral and biographical import. The greenery of their rear gardens joins together to form a beautiful circle of trees, interrupted only by the road linking the plaza with the highway. Along this main route connecting the village with the others, one finds the church, the school, and the inn.[22]

The density of population in certain big cities, notably certain neighborhoods of Paris, has reached a level of over a thousand inhabitants per hectare. Prague is even more crowded. The swelling human population seems to have reached its greatest concentration in New York, which in 1896 had a density of 1860 persons per hectare over a total area of 130 hectares.[23] Except where the military engineers have created zones around cities where dwellings are prohibited, the countryside is covered with houses and villas. In addition, the farmers are drawn toward their natural center, moving in ever closer to what is now a continuous mass of urban development and creating in the surrounding area a ring of dense population. Left with diminishing space for their fields and farmhouses, they are forced into more and more intensive labor. Shepherds become farmers, and farmers in turn become gardeners. Demographic maps show clearly the progression of this phenomenon, in which one finds an annular distribution of rural population turning to horticulture. Thus the city of Bayreuth is encircled by a zone with a population density of 109 persons per square kilometer. Around Bamberg, the density reaches 180, even though the terrain onto which this mass of people is crowded was originally of little value. As a mixture of sand and peat, it was only suitable for growing conifers; nevertheless, it has been transformed into garden soil of unsurpassed quality.[24] In the Mediterranean region, one finds that the love of the city does not so much increase the population of the countryside around the cities as depopulate it. The great privilege of participating in the discussion of the public interest has traditionally turned everyone into a city-dweller. The appeal of the agora, as in Greece, and of municipal life, as in Italy, draws the inhabitants toward the central square, where the affairs of the community are discussed, more often along the public walkways than in the resounding chambers of the city hall. Accordingly, in Provence the small landowner, rather than living among his fields, remains an inveterate city-dweller. Though he might even own a farmhouse or a country house, he refuses to live on his rural estate, but rather resides in the city, from which he can go for an outing to visit his fruit trees and do the picking. The work in the countryside is for him a secondary concern.[25]

It is quite natural that many should react against the awful swallowing up of people, the wholesale degradation of character, and the widespread corruption of the naïve souls who brew in the "infernal vat." Accordingly, some reformers call for the destruction of cities and the voluntary return of the entire population to the countryside. In an enlightened society that resolutely wills a renaissance of humanity by means of a life in the open country, such a revolution, the likes of which have never been seen before, would surely be a real possibility. If we estimate the area of the habitable lands that are pleasant and healthy at only one hundred million square kilometers, then two houses per square kilometer, with seven or eight occupants in each, would be adequate to house all of humanity. However, human nature, whose first law is sociability, would never adapt to such a dispersion. Certainly, we need the rustling of trees and the babbling of brooks, but we also require association with other people and, indeed, with all people. The entire globe becomes for humanity a great city that alone can satisfy us.

It cannot be assumed that today's immense agglomerations of structures have reached the greatest expansion imaginable. The truth is quite to the contrary. In countries of recent colonization, where people group together spontaneously according to modern interests and tastes, cities have much greater populations proportionally than those of the old countries of Europe. Some of the large centers of growth have as much as a quarter, a third, or even half the population of the entire country. In relation to the area from which it draws its population, Melbourne is a larger city than London because the surrounding population is more mobile and because it has not been necessary, as in England, to tear it away from the countryside in which it was rooted for centuries. However, this unusual concentration of population found in Australian cities stems to a large degree from the division of the land in the countryside into vast estates in which the immigrants were unable to find a place. They were driven from the *latifundia* toward the capitals.[26] In any case, the process of transplantation becomes progressively easier, and London will be able to continue its growth with a smaller expenditure of energy. At the beginning of the twentieth century, that city has only a seventh of the population of the British Isles. It is not at all impossible that some day it will have a third or a fourth of the inhabitants, especially since London is not only the center of attraction in Great Britain and Ireland, but also the most important commercial center in Europe and a large part of the colonial world. We should not be surprised at the imminent development of urban agglomerations of ten to twenty million inhabitants in the lower Thames valley, at the mouth of the Hudson, or in other centers of attraction. Indeed, we should prepare ourselves to accept such phenomena as a normal part of social life. The growth of great foci of attraction cannot be checked until an equilibrium is established between the force of attraction of the various centers on the inhabitants of the intermediate spaces. But the movement will certainly not stop then. It will be transformed more and more into a constant exchange of population between cities, a phenomenon that can already be observed and that can be compared to the circulation of the blood in the human body. There is no doubt that this new mode of functioning will give birth to new organisms, and cities, which have already been renewed so many times, will be reborn again with a new character that will correspond to the whole of social and economic evolution.

15

The Modern State (1905)

Reclus' most extended critique of the state appears in *L'Evolution, la révolution et l'idéal anarchique* and in the chapter "L'État moderne," in volume 6 of *L'Homme et la Terre* (Paris: Librairie Universelle, 1905–8), 171–223. The following text consists of the most important sections of that chapter (171–77, 188–94, and 214–23).

The world is very close to unification. All lands, including even the small islands scattered across the vast ocean, have entered into the field of attraction of one common culture, in which the European type predominates. Only in a few rare enclaves—in lands of caves where men flee the light, or in very secluded places protected by walls of rock, forests, or marshlands—have some tribes been able to remain completely isolated, living their lives outside the rhythm of the great universal life. However, as jealously as these peoples have hidden themselves, forming small, self-sufficient hereditary circles, scientific researchers have discovered them and integrated them into the whole of humanity by studying their forms, their ways of life, and their traditions, and by placing them in a social classification of which they were previously an unknown member.

The instinctive tendency of all nations to take part in the common affairs of the entire world already manifests itself in many instances in contemporary history. For example, in 1897 we witnessed the six greatest European powers (whatever their secret motives may have been) claiming to seek to maintain a balance of power in Europe, while satisfying both Turkey and Greece.[1] In the process, they fired on some unfortunate Cretans—their "brothers in Christ"—in the name of "public order." Despite the disheartening spectacle of a large deployment of force against a small people who asked only that justice be rendered to them, it was nevertheless a completely new and telling political phenomenon that soldiers and sailors of various languages and nationalities could join together, grouped in allied detachments under the orders of a leader chosen by lot among the British, Austrians, Italians, French, and Russians. This was an event with an international character, unprecedented in history because of the methodical precision with which it was carried out. It was proven that Europe as a whole is now indeed a sort of republic of states, united through class solidarity. The financial caste that rules from Moscow to Liverpool causes governments and armies to act with perfect discipline.

Since then, history has offered other examples of this council of nations that forms spontaneously in all grave political situations. Since the interests of all are at stake, each wants to take part in the deliberations and profit from the settlement. In China, for example, the temporary alliance that has been achieved between nations is strong enough to unite the military representatives of all the states in a common task of destruction and massacre. Elsewhere, notably in Morocco, the collective machinations are limited for the time being to diplomatic talks, but at any rate, the case is clear. States have an acute awareness of the effects of all events throughout the world on their own destiny, and they do their best to cope with changes in the balance of power. Nevertheless, it is very important to stress the difference between the solidarity of conservative states and that of peoples during periods of revolution, in which an upsurge takes place in the opposite direction. Whereas the year 1848 rocked the world with tremors of liberty, fifty years later we find that England hands itself over to representatives of the aristocracy and throws itself into a long war behind a band of crooks. France grapples with a recrudescence of a clerical and military mentality. Spain reestablishes the practices of the Inquisition. America, populated by immigrants, tries to close its ports to foreigners. And Turkey takes revenge against Greece.

A movement of convergence toward mutual understanding is occurring all over the world. We may therefore be permitted, in order to comprehend the transformations that will occur in the future, to take as our starting point the state of mind and practice exhibited by the civilized peoples of Europe in the management of their societies and the realization of their ideal. Obviously, each group of men moving toward the same goal will not slavishly follow the same road. It will take, according to the position that it occupies at any given time, the path that is determined by the sum total of all the individual wills that it contains. So what we propose is a kind of average that is related to the particular situation of each nation and each social element according to the temporal and spatial milieu. But in such a study, the researcher must carefully distance himself from any tendency toward patriotism, that vestige of the ancient delusion that one's nation is specially chosen by Divine Providence for the acquisition of wealth and the accomplishment of great things. Corresponding to this natural delusion of all peoples that they rank first in merit and genius is another, which Ludwig Gumplowicz called "acrochronism." Its effect is that one is content to suppose that contemporary civilization, as imperfect as it may be, is nevertheless the culminating state of humanity, and that by comparison, all past ages were barbaric. This is a "chronocentric" egoism, analogous to the "ethnocentric" egoism of patriotism.

The "rights of man" were proclaimed for thousands of years by isolated individuals and more than a century ago by an assembly that has drawn the attention of peoples ever since. Yet in present-day society these rights are still only recognized in principle, like a simple word whose meaning one hardly begins to fathom. The brutal fact of authority endures against rights, in the family and in society as well as in the state. It endures while at the same time accepting its opposite and intermingling with it in a thousand illogical and bizarre combinations. There are now very few fanatical defenders of the kind of absolute authority that gives to the prince the right of life and death over his subjects, and to the husband and father the same rights over his wife and children. Yet public opinion on such matters wavers indecisively, guided less by reason than by one's individual circumstances and personal sympathies, and by the nature of the stories one hears. Generally speaking, it can be said that man measures the strictness of his principles of liberty by his share of personal benefits from the outcome. He is absolutely strict when it is a question of events that occur on the other side of the world. But when it is a question of his own country or caste, he compromises slightly by mixing his mania for authority with conceptions of human rights. Finally, when he is directly affected, he is likely to let himself be blinded by passion, and he will gladly make authoritarian pronouncements.

In certain countries—France, for example—is it not an established custom, so to speak, that the husband has the right to kill his unfaithful wife? It is above all within the family, in a man's daily relationships with those close to him, that one can best judge him. If he absolutely respects the liberty of his wife, if the rights and the dignity of his sons and daughters are as precious to him as his own, then he proves himself worthy of entering the assembly of free citizens. If not, he is still a slave, since he is a tyrant.

It has often been repeated that the family unit is the primordial cell of humanity. This is only relatively true, for two men who meet and strike up a friendship, a band (even among animals) that forms to hunt or fish, a concert of voices or instruments that join in unison, an association to realize ideas through common action—all constitute original groupings in the great global society. Nevertheless, it is certain that familial associations, whether manifested in polygyny, polyandry, monogamy, or free unions, exercise a direct influence on the form of the state through the effects of their ethics. What one sees on a large scale parallels what one sees on a small scale. The authority that prevails in government corresponds to that which holds sway in families, though ordinarily in lesser proportions, for the government is incapable of pressuring widely dispersed individuals in the way that one spouse can pressure the other who lives under the same roof.

Just as familial practices naturally harden into "principles" for all those involved, so government takes on the form of distinct political bodies encompassing various segments of the human race that are separated from one another. The causes of this separation vary and intermingle. In one place, a difference in language has demarcated two groups. In another, economic conditions arising from a specific soil, particular products, or diverging historical paths have created the boundaries that divide them. Then, on top of all the primary causes, whether arising from nature or from stages of social evolution, is added a layer of conflicts that every authoritarian society always produces. Thus through the ceaseless interplay of interests, ambitions, and forces of attraction and repulsion, states become demarcated. Despite their constant vicissitudes, these entities claim to have a sort of collective personality and demand from those under their jurisdiction that peculiar feeling of love, devotion, and sacrifice called "patriotism." But should a conqueror pass through and erase the existing borders, the subjects must, by order of that authority, modify their feelings and reorient themselves in relation to the new sun around which they now revolve.

Just as property is the right of use and abuse, so is authority the right to command rightly or wrongly. This is understood well by the masters and also by the governed, whether they slavishly obey or feel the spirit of rebellion awakening. Philosophers have viewed authority quite differently. Desiring to give this word a meaning closer to its original one, which implied something like creation, they tell us that authority resides in anyone who teaches someone else something useful, and that it applies to everyone from the most celebrated scholar to the humblest mother.[2] Still, none of them goes so far as to consider the revolutionary who stands up to power as the true representative of authority.

Individuals and classes with power at their disposal—whether chiefs of state or aristocratic, religious, or bourgeois masters—willingly intervene with brutal force to suppress all popular initiative. In their childish and barbaric illusion, they think themselves capable of stopping the overflowing vitality of the masses, and of immobilizing society for their personal profit. But they can only lift a faltering hand. The unchanging laws of history are beginning to be understood well enough so that even the more audacious exploiters of society do not dare to run head-on into its movement. They must proceed with science and skill in order to divert it onto side roads, like a train that is switched from the main track. Up to the present, the most frequently used means—and one that unfortunately benefits most the masters of the people—consists of transforming all the energies of a nation into a rage against the foreigner. The pretexts are easy to find, since the interests of states remain different and in conflict through the very fact of their separation into distinct artificial organisms. Beyond the pretexts, there exist the memories of actual wrongs, massacres, and crimes of all sorts committed in former wars. The call for revenge still resounds, and when a new war will have passed like the terrible flames of a fire devouring everything in its path, it will also leave the memory of hatred and serve as leaven for future conflicts. How many examples one could cite of such diversions! Those in power respond to the internal problems of the government through external wars. If the wars are triumphant and the masters take advantage of the opportunity to profit from them through the consolidation of their regime, they will have debased their people through the foolish vanity they call glory. They will have made the people into shameful accomplices by inviting them to steal, pillage, and slaughter, and this solidarity of evil will cause the people's former demands to languish as their cups are once more filled with the red wine of hatred.

In addition to war, those who govern have at their disposal other powerful means of protecting themselves from any threat. These include corruption and demoralization through gambling and all forms of debauchery: betting, horse-racing, drinking, cafés, and nightclubs. "If they sing, they'll pay!" The depraved, debased, and self-hating no longer have the dignity necessary to impel them to revolt. Imagining they have the souls of lackeys, they do themselves justice by accepting their oppression. Thus the wars of the Republic and the burgeoning vices and depravity that succeeded the first years of the Revolution, with its ideals of austerity and virtue, were well timed to prepare the way for the imperial regime and the shameful debasement of character. However, this swing in the opposite direction was largely the result of a normal reaction on the part of society as a whole. It is natural for men to shift from one extreme to the other, in the same way that their lives alternate from activity to sleep, and from rest to work. Moreover, since a nation is composed of many classes and diverse groups, each of which has a particular evolution within the general one, historical movements with opposing tendencies collide and intersect, creating a complicated web that the historian can untangle only with great difficulty.

Thus during the internal struggles of the French Revolution, the people of the Vendée certainly represented the principle of the autonomous and freely federated commune, in opposition to the central government. However, through a contradiction that they were unable to grasp due to their complete lack of education, they also became defenders of the Church, whose goal was universal authority over souls, and of the monarchy, which viewed all members of the commune as nothing but corvée labor to be taxed, or even as so much meat to be sliced up on the battlefield.[3] Through a strange naïveté that would be comic were it not so tragic, the Negros of Haiti, struggling for their freedom against the white planters, enthusiastically declared themselves to be subjects of the King; and the rebels of the Spanish colonies of the New World greeted the Catholic King of Spain with cheers! Throughout history, those who revolted against any authority almost always did so in the name of another authority, as if the ideal required nothing more than changing masters. During the time of great ferment in public opinion and of intellectual liberation that led to the revolution of 1830, those who worked for the emancipation of language and for the free study of the history of art and literature of all periods and all cultures (and not only those of Greece, Rome, and the Age of Louis XIV), and those who traced their origins back to the Middle Ages and even found ancestry among the Germans and Slavs (in a word, the "romantics"), had for the most part remained royalists and Christians. On the other hand, those who championed political liberty always did so through the classical forms of the Schoolmen, in the traditional style that is the hallmark of the Academies. When Blanqui, blackened with powder, finally laid down his rifle after the three victorious days in July, he simply said: "Down with the Romantics!"[4] The revolution had disintegrated into two elements: a political one, which aimed at toppling thrones, and a literary one, which worked for the liberation of language and the extension of its domain. Each of these groups of revolutionaries was reactionary from the standpoint of the other. And each faction was quite justified in criticizing the other's illogic, irrelevancies, absurdities, and stupidities.

The historian who studies the vicissitudes of events and tries to extract what is essential relative to progress has the most difficult problem to resolve, that of discovering the parallelogram of forces underlying the thousand conflicting impulses that collide on all sides. It is easy for him to err, and he often despairs that he is witnessing a collapse when in reality there was progress, or rather when, in the overall assessment of losses and gains, human resources have actually greatly increased.

But how long and difficult does the work of true revolution seem to those who are devoted to the ideal! For if the external forms of institutions and laws respond to the pressure of deeper changes taking place, they cannot produce those changes: a new impetus must always come from the interior. To begin with, it certainly appears that the adoption of a constitution or of laws that give official expression to the victory of that part of the nation which is demanding its rights would ensure the progress that had been achieved. Yet it is possible that the result will be precisely the opposite. While it is true that any charter or laws that are agreed to by the insurgents may sanction the liberty that has been won, it is also true that they will limit it, and therein lies the danger. They determine the precise limit at which the victors must stop, and this inevitably becomes the point of departure for a retreat. For a situation is never absolutely stationary, and if movement does not occur in the direction of progress, it will occur on the side of repression. The immediate consequence of law is to lull those who have imposed it during their temporary triumph, to drain from zealous individuals the personal energy that animated them in their victorious efforts, and to transfer it to others, to professional legislators and to conservatives—in other words, to the very enemies of all progressive change. Moreover, the people are conservative at heart, and the game of revolution does not please them for long. They accept evolution because they are not suspicious of it; since they are unaware of it, it is unlikely to arouse their displeasure. Having become legalists, the former rebels are in part satisfied. They enter the ranks of the "friends of order," and reaction regains the upper hand until the arrival of new groups of revolutionaries who are not tied to the system, and who, aided by the mistakes or follies of the government, smash another hole in the ancient edifice.

As soon as an institution is established, even if it should be only to combat flagrant abuses, it creates them anew through its very existence. It has to adapt to its bad environment, and in order to function, it must do so in a pathological way. Whereas the creators of the institution follow only noble ideals, the employees that they appoint must consider above all their remuneration and the continuation of their employment. Far from desiring the success of the endeavor, in the end their greatest desire is that the goal should never be achieved.[5] It is no longer a question of accomplishing the task, but only of the profits that it brings and the honors that it confers. For example, a commission of engineers is in charge of investigating the complaints of landowners who were displaced by the construction of the aqueduct of the Avre. It would seem very simple first to study these complaints and then to respond in all fairness. But no—they begin by taking a few years to do a general survey of the region, a task that had already been done, and done well at that. Time passes, expenses accumulate, and the complaints get worse. How often has it happened that the funds allocated for some public work are notoriously insufficient, scarcely enough to maintain the scaffolding, yet the engineers run up fees as if useful work were being accomplished? How many years were necessary for that tireless association, the Loire Navigable, to obtain the authorization to create a channel in the riverbed at its own expense by constructing relatively inexpensive groins? The state would only consider works costing millions, and twenty years later the matter would probably still be under study, like so many other projects that are vital for the intelligent use of French land.

The Law is decreed by the Parliament, which arises from the People, in whom national sovereignty resides. The freer the country, the more venerable its elected legislative body, and the more important the free examination of all the implications of liberty. And no institution is more deserving of critique than parliamentary government.

The Parliament was undeniably an instrument of progress for the nation that gave birth to it, and one can understand the admiration that Montesquieu developed through studying the functioning of the British system, which is so simple, and therefore so logical. Later, during the National Assembly of 1789 and the Convention, the Parliament passed through its heroic period in France, and on the whole, played a rather positive role in the history of the gradual liberation of the individual. Since then, it has spread to all countries of the world, including the Negro republics of Haiti, Santo Domingo, and Liberia. Only Russia (1905), Turkey, China, the European colonies of exploitation, and a few other states remain without national representation. The institution has become diversified in different countries, demonstrating shortcomings in some cases and strengths in others, but one finds everywhere a profound divergence between the evolution of a people and that of its legislative body.

Even if one sets aside systems with poll taxes and plural voting, ignores the fact that with rare exceptions the feminine half of the population is not "represented" at all, and considers only universal suffrage that is honestly applied, one still cannot claim that the laws voted on by the majority of the elected representatives, who are themselves selected by the majority of the voters, express the opinion of the majority of electors. In fact, the opposite is often true. This defect, which is purely mathematical, might be negligible if the state contained only two factions, since the losses and gains would balance out on the whole, but it becomes so much more serious as life intensifies and opinions become more diverse. Yet the Swiss are alone in conferring on the entire electorate the final adoption or rejection of each new law.

Except in very rare cases, the spectacle presented by countries during an election would hardly delight a man of principles. Whether an electoral committee drafts the candidate, or whether he violates his own modesty, ambitions inevitably emerge, and machinations, extravagant promises, and lies have free rein. Moreover, it is certainly not the most honest candidate who has the best chance of winning. Since the legislators must be knowledgeable about all sorts of problems—local and global, financial and educational, technical and moral—no particular ability recommends the candidate to the voters. The winner may owe his success to a certain provincial popularity, his good-natured qualities, his oratorical skills, or his organizational talents, but frequently he is also indebted to his wealth, his family connections, or even the terror that he can inspire as a great industrialist or large property owner. Most often, he will be a man of the party; he will be asked neither to involve himself in public works, nor to facilitate human relations, but rather to fight against one faction or another. In short, the composition of the legislature does not at all reflect that of the nation. It will be generally inferior in moral qualities, since it is dominated by professional politicians.

Once elected, the representative is in fact independent of his electors. It is left up to him to decide on the thousand issues of each day according to his own conscience, and if he does not take the side of his constituents, there exists no recourse against his vote. Far from having any accountability during the four, seven, or nine years of his mandate, and well aware that he can now commit crimes with impunity, the elected official finds himself immediately exposed to all sorts of seductions on behalf of the ruling classes. The newcomer is initiated into the legislative traditions under the leadership of the veteran parliamentarians, adopts the *esprit de corps*, and is solicited by big industry, high officials, and above all, international finance. Even if the parliament happens to be composed of a majority of honest people, it develops a peculiar mentality based entirely on negotiations, compromises, recantations, dealings that must not reach the ears of the general public, and bargaining in the corridors that is covered up by brilliant jousting between skilled orators. All noble character is debased, all sincere conviction contaminated, and all honest intention destroyed.

Thus it is not surprising that so many men refuse to help sustain such an environment by means of their vote and to cooperate in the "conquest of state power." The revolutionaries at least realize that the forms of the past will endure as long as the workers support their existence and compromise with them, even if only to modify them. They can only deplore the naïveté of those who think that they can "make the Revolution armed to the teeth with ballots." In order to maintain this illusion, one must ignore the real weakness of this allegedly sovereign parliament, closing one's eyes to the far more powerful institutions that gather around it, playing with it like a cat with a mouse.

All the movements for emancipation stand together, although the insurgents are often unaware of each other, and they even hold on to their atavistic enmities and resentments. From England and Germany to France and Italy, there are many workers who despise one another, though this does not prevent them from helping each another in their common struggle against capitalist oppression. Similarly, among the women who have thrown themselves impetuously into the battle for equality between the sexes, there were at first a very significant number who, with their rather patrician or high-brow tendencies, harbored a pious disdain of the worker in his worn-out or dirty clothes. Nevertheless, since the early days of "feminism," we have witnessed the heroism of brave women who go to the prostitutes to join them in solidarity to protest the abominable treatment to which they have been subjected, and the shocking bias of the law in favor of the corrupters and against their victims. Risking insults and the most unsavory contacts, they dared to enter the brothels and form an alliance with their scorned sisters against the shameful injustice of society. Consequently, the coarse laughter and vulgar insults that greeted their first steps gave way to a profound admiration on the part of many who had mocked them. Here is a courage of a different order than that of the fierce soldier who, seized with a bestial fury, lunges with his sword or fires his rifle.

Obviously, all of the claims of women against men are just: the demands of the female worker who is not paid as much as the male worker for the same labor, the demands of the wife who is punished for "crimes" that are mere "peccadilloes" when committed by the husband, and the demands of the female citizen who is barred from all overt political action, who obeys laws that she has not helped to create, and who pays taxes to which she has not consented. She has an absolute right to recrimination, and the women who occasionally take revenge are not to be condemned, since the greatest wrongs are those committed by the privileged. But ordinarily, a woman does not avenge herself at all. To the contrary, at her conventions she naïvely petitions legislators and high officials, waiting for salvation through their deliberations and decrees; however, experience teaches women year after year that freedom does not come begging, but rather must be conquered. It teaches them, moreover, that in reality their cause merges with that of all oppressed people, whoever they may be. Women will need to occupy themselves henceforth with all people who are wronged, and not only with the unfortunate women forced by poverty to sell their bodies. Once all are united, all the voices of the weak and the downtrodden will thunder with a tremendous outcry that will indeed have to be heard.

Make no mistake about it. Those who seek justice would have neither a chance of realizing it in the future nor a single ray of hope to console them in their misery if the league of all enemy classes had no defections and remained as solid as the human wall of an infantry formation. However, countless renegades leave their ranks. Some go without hesitation to augment the camp of the rebels, while others disperse here and there, somewhere between the ranks of the innovators and the conservatives. In any case, they are too far from their original position to be brought back at the moment of battle. It is perfectly natural that organized bodies are thus weakened by a loss of their best elements through a continual migration. The study of the interconnected facts and laws revealed by contemporary science, the rapid transformation of society, new conditions in the environment, and the need for mental balance in those who are logically attracted to the search for truth—all this creates for the young a milieu completely different from that entailed by a traditional society with its slow and painful evolution. It is true that the representatives of ancient monopolies also gain recruits, especially among those who, tired of suffering for their ideas, finally want to try out the joys and privileges of this world, to eat when they are hungry and take their turn living as parasites. But whatever the particular worth of a given individual who changes his ideals and practices, it is certain that the revolutionary offensive benefits by this exchange of men. It receives those who have conviction and determination, young people with boldness and will, whereas those whom life has defeated head for the camp of the parties of reaction and bring with them their discouragement and their faintheartedness.

The state and the various elements that constitute it have the great disadvantage of acting according to a mechanism so regular and so ponderous that it is impossible for them to modify their movements and adapt to new realities. Not only does bureaucracy not assist in the economic workings of society, but it is doubly harmful to it. First, it impedes individual initiative in every way and even prevents its emergence; second, it delays, halts, and immobilizes the works that are entrusted to it. The cogs of the administrative machine work precisely in the opposite direction from those functioning in an industrial establishment. The latter strives to reduce the number of useless articles, and to produce the greatest possible results with the simplest mechanism. By contrast, the administrative hierarchy does its utmost to multiply the number of employees and subordinates, directors, auditors, and inspectors. Work becomes so complicated as to be impossible. As soon as business arises that is outside the normal routine, the administration is as disturbed as a company of frogs would be if a stone were thrown into their swamp. Everything becomes a pretext for a delay or a reprimand. One withholds his signature because he is jealous of a rival who might benefit from it; another because he fears the displeasure of a supervisor; a third holds back his opinion in order to give the impression of importance. Then there are the indifferent and the lazy. Weather, accidents, and misunderstandings are all used as excuses for the results of ill will. Finally, files disappear under a layer of dust in the office of some malevolent or lazy manager. Useless formalities and sometimes the physical impossibility of providing all of the desired signatures halts business, which gets lost like a parcel en route between capitals.

The most urgent projects cannot be accomplished because the sheer force of inertia of the bureaucracy remains insurmountable. This is the case with the island of Ré, which is in danger of some day being split in two by a storm. On the ocean side, it has already lost a strip of land several kilometers wide in some places, and currently all that remains at the most threatened point is an isthmus of less than one hundred meters. The row of dunes that forms the backbone of the island is very weak there. Considering all the facts, it is inevitable that one day, during a strong equinoctial tide, a raging westerly wind will push the waves across the peduncle of sand and open up a large strait through the swamps and fields. Everyone agrees that it is urgent to construct a strong seawall at the weak point on the island; however, some time ago a small fort was built, a worthless construction now abandoned to the bats, without even a man garrisoned there. No matter, it is in principle under the supervision of the corps of engineers, and consequently all public works are necessarily halted in its vicinity. This part of thc island will have to perish. Not far from there, the waters of a gulf have intruded into the salt marshes and changed them into a shallow estuary. It would be easy to recover these "Lost Marshes," and the surrounding residents have formulated a proposal to do so. But the invasion of the sea has made state property of the area, and the series of formalities that the recovery of the land would entail seems so interminable that the undertaking has become impossible. The lost land will remain lost unless a revolution abolishes all clumsy intervention from an ignorant and indifferent state and restores the free management of interests to the interested parties themselves.

In certain respects, minor officials exercise their power more absolutely than persons of high rank, who are by their very importance constrained by a certain propriety. They are bound to respect social decorum and to conceal their insolence, and this sometimes succeeds in soothing them and calming them down. In addition, the brutalities, crimes, or misdemeanors committed by important figures engage everyone's attention. The public becomes enthralled with their acts and discusses them passionately. Often they even risk being removed from office through the intervention of deliberative bodies and bringing their superiors down with them. But the petty official need not have the slightest fear of being held responsible in this way so long as he is shielded by a powerful boss. In this case, all upper-level administration, including ministers and even the king, will vouch for his irreproachable conduct. The uncouth can give free rein to crass behavior, the violent lash out as they please, and the cruel enjoy torturing at their leisure. What a hellish life it is to endure the hatred of a drill sergeant, a jailer, or the warden of a chain gang! Sanctioned by law, rules, tradition, and the indulgence of his superiors, the tyrant becomes judge, jury, and executioner. Of course, while giving vent to his anger, he is always supposed to have dispensed infallible justice in all its splendor. And when cruel fate has made him the satrap of some distant colony, who will be able to oppose his caprice? He joins the ranks of kings and gods.

The arrogant, do-nothing petty bureaucrat who, protected by a metal grating, can take the liberty of being rude toward anyone; the judge who exercises his "wit" at the expense of the accused he is about to condemn; the police who brutally round up people or beat demonstrators; plus a thousand other arrogant manifestations of authority—this is what maintains the animosity between the government and the governed. And it must be noted that these daily acts do not wrap themselves in the mantle of the law but rather hide behind decrees, memos, reports, regulations, and orders from the prefect and other officials. The law can be harsh and indeed unjust, but the worker crosses its path only rarely. In certain circumstances, he can even go through life without suspecting that he is subject to it, as when he is unaware that he is paying some tax. But every time he acts, he is confronted with decisions decreed by officials whose irresponsibility differs from that of the members of parliament. The decisions of the former are without recourse and continually remind the individual of the guardianship that the state exercises over him.

The number of high and low officials will naturally grow considerably, in proportion to increases in budgetary resources and to the extent that the treasury contrives to find new means of extracting additional revenues from whatever may be taxed. But the proliferation of employees and staff members results above all from what we like to call "democracy," that is, from the participation of the masses in the prerogatives of power. Each citizen wants his scrap, and the main preoccupation of those who already have an official post is to classify, study, and annotate the applications of others who seek a position. The budget has paid for, and possibly continues to pay for, a forest ranger on the island of Ouessant, which has a grand total of eight trees—five in the garden of the curé and three in the cemetery!

So much pressure is exerted on the government by the multitude of supplicants that the acquisition of distant colonies is due in very large part to the concern for the distribution of government positions. One can judge the so-called colonization of many countries by the fact that in Algeria in 1896 there were a little more than 260,000 French residing within the territorial boundaries, of which more than 51,000 were officials of all kinds. This constitutes roughly a fifth of the colonists,[6] yet one must also take into account the 50,000 soldiers stationed there. This brings to mind the inscription added on a map to the name of the "town" of Ushuaia, the southernmost urban settlement of the Americas and of the world: "Seventy-eight inhabitants, all officials"!

France is an example of such a "democratization" of the state since it is managed by approximately six hundred thousand participants in the exercise of sovereign power. But if one adds to the officials in the strict sense those who consider themselves as such, and who are indeed invested with certain local or temporary powers, as well as those distinguished from the mass of the nation through titles or distinguishing marks, such as the village policemen and the town criers, not to mention the recipients of decorations and medals, it becomes apparent that there are more officials than soldiers. Moreover, the former are, as a group, much more energetic supporters of the government that pays them. Whereas the soldier obeys orders out of fear, the official's motivation stems not only from forced obedience but also from conviction. Being himself a part of the government, he expresses its spirit in his whole manner of thinking and in his ambitions. He represents the state in his own person. Moreover, the vast army of bureaucrats in office has a reserve force of a still greater army of all the candidates for offices, supplicants and beggars of favors, friends, and relations. Just as the rich depend on the broad masses of the poor and starving, who are similar to them in their appetites and their love of lucre, so do the masses, who are oppressed, persecuted, and abused by state employees of all sorts, support the state indirectly, since they are composed of individuals who are each preoccupied with soliciting jobs.

Naturally, this unlimited expansion of power, this minute allocation of positions, honors, and meager rewards, to the point of ridiculous salaries and the mere possibility of future remuneration, has two consequences with opposing implications. On the one hand, the ambition to govern becomes widespread, even universal, so that the natural tendency of the ordinary citizen is to participate in the management of public affairs. Millions of men feel a solidarity in the maintenance of the state, which is their property, their affair. At the same time, the growing debt of the government, divided into thousands of small entitlements to income, finds as many champions as it has creditors drawing the value of their income coupons from quarter to quarter. On the other hand, this state, divided into innumerable fragments, showering privileges on one or another individual whom all know and have no particular reason to admire or fear, but whom they may even despise—this banal government, being all too well understood, no longer dominates the multitudes through the impression of terrifying majesty that once belonged to masters who were all but invisible and who only appeared before the public surrounded by judges, attendants, and executioners. Not only does the state no longer inspire mysterious and sacred fear, it even provokes laughter and contempt. It is through the satirical newspapers, and especially through the marvelous caricatures that have become one of the most remarkable forms of contemporary art, that future historians will have to study the public spirit during the period beginning with the second half of the nineteenth century. The state perishes and is neutralized through its very dissemination. Just when all possess it, it has virtually ceased to exist, and is no more than a shadow of itself.

Institutions thus disappear at the moment when they seem to triumph. The state has branched out everywhere; however, an opposing force also appears everywhere. While it was once considered inconsequential and was unaware of itself, it is constantly growing and henceforth will be conscious of the work that it has to accomplish. This force is the liberty of the human person, which, after having been spontaneously exercised by many primitive tribes, was proclaimed by the philosophers and successively demanded with varying degrees of consciousness and will by countless rebels. Presently, the number of rebels is multiplying, and their propaganda is taking on a character that is less emotional than it was previously and much more scientific. They enter the struggle more convinced, more daring, and more confident of their strength, and they find an environment that offers more opportunities to avoid the grip of the state. Here is the great revolution that is developing and even reaching partial fulfillment before our eyes. In the past, society has functioned through distinct nations, separated by borders and living under the domination of individuals and classes who claim superiority over other men. We now see another mode of general evolution that intermingles with the previous one and begins to replace it in an increasingly regular and decisive manner. This mode consists of direct action through the freely expressed will of men who join together in a clearly defined endeavor, without concern for boundaries between classes and countries. Each accomplishment that is thus realized without the intervention of official bosses and outside the state, whose cumbersome machinery and obsolete practices do not lend themselves to the normal course of life, is an example that can be used for larger undertakings. Erstwhile subjects become partners joining together in complete independence, according to their personal affinities and their relation to the climate that bathes them and the soil that supports them. They learn to escape from the leading strings that had guided them so badly, being in the hands of degenerate and foolish men. It is through the phenomena of human activity in the arenas of labor, agriculture, industry, commerce, study, education, and discovery that subjugated peoples gradually succeed in liberating themselves and in gaining complete possession of that individual initiative without which no progress can ever take place.

16

Culture and Property (1905)

Some of Reclus' most extensive comments on historical forms of property are found in "Culture and Property," which is in volume 6 of *L'Homme et la Terre* (Paris: Librairie Universelle, 1905–8), 225–311. There he discusses the differences between large and small property holdings, individual and communal property, and cooperative and competitive practices. The following selections are taken from that chapter (268–71, 280–85). The text includes some of Reclus' most eloquent encomiums to cooperation and stinging criticisms of concentrated economic power.

There is not a single European country in which the traditions of the old communal property have entirely disappeared. In certain areas, notably in the Ardennes and in the steep mountainous regions of Switzerland, where the peasants did not have to submit to the kind of oppression to which the German villagers were subjected after the wars of the Reformation, communal property is still widespread enough to constitute a considerable part of the territory.

In the Belgian Ardennes, the collective lands are composed of three parts: the woods, the freshly cleared ground [*sart*], and the pastures. They also often include arable land and quarries. The woods, which form the largest part of the property, are divided into a certain number of sections, generally twenty to twenty-two. Each year, one section is divided by drawing lots among the various households of the commune, the bark of the oaks having been previously stripped for the benefit of the communal coffers. For the work with heavy wood, the families divide into groups of five, whose members rotate the responsibility of cutting down the trees, squaring the timber, and transporting it. After the cutting, each person proceeds to clear the portion of the land that fell to his lot and sows the rye that he will harvest the following year. Two and a half years after harvesting the rye, the inhabitants apportion the broom plants that have grown in the clearings, after which the section, in which new growth has already begun, is left to itself until the same operations recommence. The grazing is communal and without any special organization, and takes place on the uncultivated lands, in the mature woods, and in the brush six or seven years after a cutting. Stones may be quarried freely, barring any previous notice to the contrary.

These customs clearly influence the moral character of individuals and greatly develop their spirit of solidarity, mutual kindness, and heartfelt friendliness. Thus it is customary to form voluntary work crews for the benefit of those who need work done. The latter need only to state their request by proceeding noisily through the village, calling out, "So-and-so needs something done! Who wants to help out?" Immediately a group appears and its members put their heads together to figure out who can best undertake the job, and the service is rendered.[1] Such stories also come to us from the Queyras.[2]

In all of Switzerland, two-thirds of the alpine prairies and forests belong to the communes, which also own peat bogs, reed marshes, and quarries, as well as fields, orchards, and vineyards. On many occasions when the co-proprietors of the commune have to work together, they feel as though they are at a festival rather than at work. The young men and women climb to the high mountain pastures, driving their herds before them to the harmonious clinking of the bells. At other times, the work is more difficult. While the snow still covers the ground, the woodsmen, armed with axes, cut the high pines in the communal forest. They strip the sawlogs and slide them down the avalanche corridors to the torrent that will carry them away in its bends and rapids.

Then there are the evening gatherings on winter nights, in which all are summoned to the home of whoever has the most urgent work, whether it is to shell corn, hull nuts, or make wedding gifts for a woman engaged to be married. During these gatherings, the work is a pleasure. The children want to participate, for everything is new to them. Instead of going to bed, they stay up with the adults and are given the best of the chestnuts roasting under the hot embers. When dreamtime is near, they listen to songs and are told stories, adventures, and fables, which are transformed by their imaginations into marvelous apparitions. It is often during such nights of mutual good will that a child's being permanently takes shape. Here, one's loves in life are kindled, and life's bitterness is made sweeter.

Thus the spirit of full association has by no means disappeared in the communes, despite all the ill will of the rich and the state, who have every interest in breaking apart these tightly bound bundles of resistance to their greed or power and who attempt to reduce society to a collection of isolated individuals. Traditional mutual aid occurs even among people of different languages and nations. In Switzerland, it is customary to exchange children from family to family, between the German and the French cantons. Similarly, the country people of Béarn send their children to the Basque country, welcoming in turn young Basques as farm boys. In this way, they will all soon learn the two languages without the parents having to spend any money. Finally, all individuals with a similar trade and common interests—whether they be coal merchants, hunters, or sailors—have established virtual confraternities having neither written constitutions nor signatures, but nevertheless forming small, close-knit republics. Throughout the world, carnival performers who meet by chance on the road are allied in a sort of freemasonry that is far more solemn than that of the "brothers" who gather in the temples of Hiram.[3]

It is evident that anyone who becomes master over his fellow man through war, conquest, usury, or any other means thereby establishes private property for his own advantage. For by appropriating the man, he also takes possession of another's labor and of the product of that labor, and finally of that portion of the common soil on which his slave produces crops. No matter how tenaciously the people may have sought to maintain their ancient traditions, the power of kings has inevitably led these rulers to indulge their caprice. They take men and land, and dispense all according to their whims. The forms of gratitude, the homage of vassals, and the circumstances of tenure have varied according to the country and the age, but the essential fact is that ownership of the land was no longer secured for those who worked it but was instead granted to one who was incapable of handling a spade or driving a plow.

Just as common property and private property conflict, there is a constantly raging battle between large and small property. Not only does each create class groupings hostile to one another, but they also collide as two different and enemy systems. Although each arises from the appetites and passions of man, the two forms of property are presented by their advocates as systems that should be maintained permanently because of their essential virtues. First of all, small ownership, which seems closer to natural equity, is vaunted as the ideal state. It offers to the farming family a life of constant work and regular employment to fill its hours and days. Even when the fields are fallow, the members of the household must tend to the livestock and prepare their produce. They also decorate their homes, and in this way art plays a normal role in the life of the peasant. Novelists delight in the rustic cottage, which becomes the charming setting for the idyll of their dreams. But though the dream has been realized many times, it is much more likely that a wretched poverty will inhabit the hearth. And even if a humble family is lucky enough to enjoy modest comfort, what can they do to enlarge their horizons, to expand their ideas, to renew their intellectual resources, or even to increase their knowledge of their own industry? The routine that binds them to the hereditary soil also holds them tightly in the grip of the customs of the past. However free they may appear to be, they nevertheless possess the souls of slaves.

The owners of vast landholdings claim to be educators in the science of agriculture in order to justify the usurpation of communal and private lands due to their birth, hereditary wealth, or speculations. This claim is particularly inappropriate in the case of those powerful lords who are careful to live somewhere other than on their own lands, like most of the nobility of Irish estates, who are well aware of the hatred their tenant farmers feel for them. Is it not, then, simply ludicrous to speak of them as "educators" of any sort? And what about those who might otherwise be warmly received by serfs reconciled to the condition of non-ownership, but who, concerned only with receiving their income, hand over the entire burden of management to stewards, trustees, or lawyers, for whom the management of the estate is also far from being a selfless duty?

It is true that in certain countries renowned agronomists owning large estates have instituted excellent methods of cultivating the soil, managed their fields as scientifically as the chemical industries that utilize the most up-to-date processes, introduced new species of plants and animals, and adopted practices that were previously unknown. One must not forget, however, that the *latifundium*[4] in its essence inevitably requires that the vast majority be deprived of land. If a few have much, it is because the majority no longer have any. Some large owners are seized with a hunger for land and also desire to be admired as local benefactors. But the devouring of the surrounding land by the large estates is hardly less disastrous than fire and other devastations. Moreover, it produces the same end result, which is the ruin not only of populations but also frequently of the land itself. Intelligent large landholders can no doubt train excellent farm hands, and they will certainly have domestics of impeccable correctness. But even assuming that the productive industry initiated by them provides more than enough labor for the entire local population, is it not inevitable that their authoritarian and absolutist manner of regimenting labor will create subjects rather than produce dignified equals? They make every effort to preserve the essentially monarchical character of society. Moreover, they try to return to the past by destroying all democratic elements in their milieu in order to reconstitute a feudal world where power belongs to those they deem to be the most deserving—that is to say, to themselves. And whether or not they are the most deserving, they remain the most privileged. One need only study a map of France to verify the influence exerted by large estates. Among the reasons that certain cantons automatically fall into the hands of reactionary representatives and masters, who are both clericalist and militarist, none is more crucial than the influence of the large landowners. They have no need to tell their flunkies and farm hands how to vote, for they easily lead them so far down the path of moral degradation that they willingly vote in favor of a regime of obedience to the traditional master. The same spirit determines the voting of lackeys and tradesmen in the elegant neighborhoods of the cities and in the resorts.

Furthermore, is it not possible that if all its effects are considered, large ownership actually produces less material improvement than does small property, as divided up as the latter may be? If, taking the economy of France as a whole, one were to make a detailed comparison of the net profit produced by large estates under individual management and the losses to the communes resulting from the parks reserved for the privileged few, the hunting grounds, and the moors that displace small property, it is quite possible that, on balance, the losses would be greater. We would discover that large land ownership is for modern peoples what it was for ancient ones—a fatal plague. Furthermore, initiative has emerged not only among rich agronomists but also—though with less ostentation and acclaim—in small holdings among truck farmers, horticulturalists, and small farmers. The poor person is certainly a slave to routine and risks his few pennies, eaten away by taxes and usury, only with extreme prudence. But risk them he does. Some know how to observe, experiment, and learn, so that over many generations and centuries they carry out experiments of long-lasting value. The case is clear: the land of the austere peasant today yields twice as much as it did when Young traveled through the provinces of France and noted its disheartening poverty.[5] Only through private initiative can there be progress, but the union of forces that enjoys all the advantages of large and small ownership has hardly begun to appear. There are only signs of its coming.

In considering the consequences of large property ownership, we must not forget the obstacles that it places in the way of free movement when the surrounding populations do not know how to bypass restrictions. In Great Britain, the "right of way" issue excites local opinion in twenty different places at any given time. The inhabitants find themselves cut off from the old roads, one after the other. Pity the communities that appeal to a court of law if they lack indisputable titles! In many districts in Scotland, landlords have forbidden by law all access to the mountains, and pedestrians are reduced to using the same roadway at the bottom of the valley as do bicycles and automobiles. The maps of the Ordnance Survey even caution that "the existence of a road on a map does not imply the right to use it." And woe to the traveler who takes it upon himself to enter the underbrush or to pass through a fallow field! The last tollgates are now disappearing—as recently as 1893, 600,000 francs were paid for the removal of a turnpike that prevented livestock from having free access to Gower Street in London (the equivalent of Rue Bergère in Paris); however, numerous new prohibitive barriers have replaced these old tollgates. The usual excuse given by the landowners for closing the roads that cross their estate is the preservation of game, so poaching becomes an inevitable corollary of large landholdings. There is a stark contrast between the hunting trophies on which the legally authorized hunter prides himself and the slaughter committed by his nocturnal counterpart as well as the fishing by dynamite, which depopulates a river in a few hours. Moreover, the legal consequences are far from the same for these two sorts of hunters. Manhunting is permitted in practice to the property owner and his guards. On the other hand, one cannot begin to estimate how many during the nineteenth century have spent years in prison or at hard labor, or have even gone to the scaffold, as a result of hunting the rabbit and the "sacred bird."

Statesmen and economists are often interested in encouraging small property ownership. In Denmark, notably, every opportunity is offered for the easy acquisition of property of less than four hectares. Another example that comes to mind is the homestead exemption found in the United States, in which a small area of land per family as well as the house that the family occupies are declared non-transferable and unseizable, with conditions that vary somewhat from state to state. But it is obvious that such a system must remain limited to a small segment of the population. Otherwise, if each producer had access to the soil, his independence would be assured, and the current conception of society would be shaken to its very foundation. Also, one can be sure that nothing like this will ever become law in France, unless restrictions are imposed to make the effects illusory. Among European peoples, the Icelanders are alone in taking precautions against the monopolization of land. Since 1884, the property owner who does not cultivate the land himself has been obliged to rent it to another.

17

Progress (1905)

"Progress" is the final chapter of Reclus' final work, *L'Homme et la Terre*. It is one of the most comprehensive statements of his view of human nature, historical development, and social values. This text is translated in its entirety from volume 6 of *L'Homme et la Terre* (Paris: Librairie Universelle, 1905–8), 501–41.

"Progress," in the strictest sense of the word, is meaningless, for the world is infinite, and in its unlimited vastness, one is always as distant from the beginning as from the end. The movement of society ultimately reduces to the movements of the individuals who are its constitutive elements. In view of this fact, we must ask what progress in itself can be determined for each of these beings whose total life span from birth to death is only a few years. Is it no more than that of a spark of light glancing off a pebble and vanishing instantly into the cold air?

The idea of progress must be understood in a much more qualified sense. The common meaning of this word has been passed down to us by the historian Gibbon, who states that "since the beginning of the world, each age has increasingly improved the material wealth, the happiness, the scientific knowledge, and perhaps the virtue of the human species."[1] This definition, which is somewhat questionable from the standpoint of moral evolution, has been adopted by modern writers and modified, expanded, or narrowed in various ways. In any case, the common view of the word "progress" is that it encompasses the general improvement of humanity throughout history. But it would be a mistake to attribute to every other epoch of life on earth an evolution analogous to that which contemporary humankind has experienced. There are quite plausible hypotheses dealing with the geological time of our planet that lend a great deal of support to the theory of a fluctuation of ages corresponding on a larger scale to the phenomenon of our alternating summers and winters. A back-and-forth motion encompassing thousands or millions of years or of centuries would result in a succession of distinct and contrasting periods in which life evolves in ways that are very different from one another. What would become of present-day humanity if there were another "great winter"—that is, if a new ice age were again to cover the British Isles and Scandinavia with a continuous sheet of ice, and our museums and libraries were to be destroyed by the severe cold? Would we simply have to hope that the two poles would not simultaneously become colder, and that man would be able to survive by gradually adapting to the new conditions and by moving the treasures of our present civilization to warmer climates? But if there were a widespread cooling, is it conceivable that an appreciable decrease in solar heat, which is the source of all life, and the gradual depletion of our energy resources, could permit continued improvement of culture or real progress? Today we are already able to confirm that the normal consequences of the drying of the earth following the ice age caused unquestionably regressive phenomena in regions of Central Asia. Dried-up rivers and lakes, and waves of invading dunes, brought with them the demise of cities, civilizations, and nations themselves. Sandy deserts replaced countryside and cities. Man was not able to hold his ground against a hostile nature.

Whatever conception we might have of progress, one point seems completely indisputable: in different epochs, certain individuals have emerged who, through some characteristic, have attained great prominence among men of all times and nations. One can think of scores of names of persons who, by their perspicacity, hard work, deep-seated goodness, moral virtue, artistic sensibility, or some other aspect of character or talent, constitute ideal and unsurpassable types in their particular sphere. The history of Greece in particular presents great examples, but other human groups have possessed them, as we have often surmised from myths and legends. Who could claim to be better than Shakyamuni, more artistic than Phidias, more inventive than Archimedes, or wiser than Marcus Aurelius? If there has been progress during the past three thousand years, it must consist of a greater diffusion of this initiative previously reserved for a few, and of a better utilization of gifted minds by society.

Some great thinkers are not satisfied with these fundamental restrictions in the concept of progress and furthermore deny that there could be any real improvement in the general state of humanity. According to them, the whole idea of progress is completely illusory and only has meaning from an individual point of view. Indeed, for most men, the fact of change is synonymous with either the idea of progress or that of regression, depending on its relative motion toward or away from the step occupied by the observer on the ladder of beings. The missionaries who encounter magnificent savages moving about freely in their nakedness believe that they will bring them "progress" by giving them dresses and shirts, shoes and hats, catechisms and Bibles, and by teaching them to chant psalms in English or Latin. And what triumphant songs in honor of progress have not been sung at the opening ceremonies of all the industrial plants with their adjoining taverns and hospitals![2] Certainly, industry brought real progress in its wake, but it is important to analyze scrupulously the details of this great evolution! The wretched populations of Lancashire and Silesia demonstrate that their histories were not a record of unadulterated progress. It is not enough to change one's circumstances and enter a new class in order to acquire a greater share of happiness. There are now millions of industrial workers, seamstresses, and servants who tearfully remember the thatched cottages of their childhoods, the outdoor dances under the ancestral tree, and the evening visits around the hearth. And what kind of "progress" is it for the people of Cameroon and of Togo to have henceforth the honor of being protected by the German flag, or for the Algerian Arabs to drink aperitifs and express themselves elegantly in Parisian slang?

The word "civilization," which is ordinarily used to indicate the progressive state of a particular nation, is, like the word "progress," one of those vague expressions that confounds various meanings. For most individuals, it characterizes only the refinement of morals and, above all, those outward conventions of courtesy that merely prevent men of awkward bearing and rude manners from claiming moral superiority over courtiers playing their elegant madrigals. Others see in civilization only the sum total of material improvements due to science and modern industry. To them, railroads, telescopes and microscopes, telegraphs and telephones, dirigibles and flying machines, and other inventions seem sufficient evidence of the collective progress of society. They do not want to know anything beyond this or to probe into the depths of the great organism of society. But those who study it from its beginnings note that each "civilized" nation is composed of superimposed classes representing in this century all successive previous centuries with their corresponding intellectual and moral cultures. Present-day society contains within itself all past societies in the form of survivals, and when seen in close juxtaposition, their vastly differing conditions of life present a striking contrast.

Obviously, the word "progress" can cause the most unfortunate misunderstandings, depending on the meaning attributed to it by those who use it. Buddhists and the exegetes of their religion could number the various definitions of nirvana in the thousands. Likewise, philosophers, according to their ideals of life, are capable of viewing the most varied (and even the most contradictory) evolutions as examples of "moving forward." There are some for whom repose is the ultimate good, and they make a vow, if not for death, at least for perfect peace of body and mind and for "order," even if this consists of no more than routine. What these weary beings consider to be Progress is certainly looked upon as something entirely different for men preferring a perilous freedom to a peaceful servitude. However, the average view of progress is identical to that of Gibbon. It entails the improvement of physical being from the standpoint of health, material enrichment, the growth of knowledge, and finally the perfection of character, which becomes distinctly less cruel, more respectful of the individual, and perhaps more noble, generous, and dedicated. From this point of view, the progress of the individual merges with that of society, united by the force of an increasingly intimate solidarity.

In view of the uncertainty concerning the meaning of progress, it is important to study each historical fact from a sufficient distance so as not to become lost in the details, and to find the necessary vantage point from which to determine the true relationships to the whole of all the interconnected civilizations and peoples. There are examples of men of high intelligence who absolutely deny not only progress but even any concept of a sustained evolution for the better. Ranke, though otherwise a historian of great value, sees in history only successive periods, each having its own peculiar character and manifesting itself through various tendencies that give a distinct, unexpected, and even "piquant"[3] life to the different tableaux of each epoch and each people.[4] According to this conception, the world appears as a sort of picture gallery. If there were progress, says the pietist writer, men would be assured of improvement from century to century, and they would therefore not be "directly dependent on the divinity," who sees all successive generations in the course of time with an impartial eye, as if their relative value were exactly equal. Ranke's opinion goes against those usually encountered since the eighteenth century and justifies once more the observation of Guyau that "the idea of progress is antagonistic to that of religion."[5] Because of the sovereign authority of gods and dogmas that lasted through the ancient and medieval ages, this idea of progress remained dormant for a long time, hardly awakened by the most open-minded philosophers of the ancient world, and came to life with full self-consciousness only with the Renaissance and the period of modern revolutions. Indeed, all religion proceeds from the principle that the universe emerged from the hands of a creator; in other words, that it had its origin in supreme perfection. As the Bible states, God looked at his work and saw that it was "good," and even "very good."[6] Following this original state marked by the seal of divinity, the movement resulting from the actions of imperfect men could only continue toward decline and fall; regression was inevitable. After the Golden Age, these creatures ended up falling into the Iron Age. They left the paradise where they had lived happily, to be engulfed by the waters of the Flood, from which they emerged only to lead thereafter an aimless life.

Moreover, the entrenched institutions of monarchy and aristocracy, and all the official and exclusive creeds founded and masoned, so to speak, by men who claimed, and indeed were even certain, that they had achieved perfection, presupposed that all revolution and all change must be a fall, a return to barbarism. These ancestors and forefathers, glorifiers of "the olden days," played a large role along with gods and kings in the denigration of the present relative to the past and in the creation of a prejudice that regression is inevitable. Children have a natural tendency to regard their parents as superior beings, and these parents have in turn done the same. Such attitudes have been successively deposited in minds like alluvial soil on the banks of a river, and have consequently created a veritable dogma of man's irremediable fall from grace. Even in our time, is it not a widespread practice to hold forth in prose or verse on "the depravity of our century"? For example, the same people who praise the "inevitable progress of humanity" speak readily of its "decline," thus showing a complete (though nearly unconscious) lack of logic. Two contrary currents intersect in their speech as well as in their views. Indeed, previously held notions collide with new ones, even among reflective persons who do not speak unthinkingly. Though the weakening of religions is interrupted by sudden revivals, they must nevertheless succumb to the force of theories that explain the formation of the world by slow evolution, the gradual emergence of things from primitive chaos. And what is this phenomenon if not by definition progress itself—whether acknowledged implicitly, as by Aristotle, or in precise, eloquent words, as by Lucretius?[7]

The idea that there has been progress during the brief span of each human generation and in the whole of human evolution owes its persuasiveness largely to geological research, which has revealed in the succession of phenomena, if not a "divine plan," as it was once called, a natural evolution that gradually refines life by means of increasingly complex organisms. Thus the first life-forms whose remains or traces can be seen in the most ancient strata of the earth present rudimentary, uniform, and scarcely differentiated features, and constitute increasingly successful sketches of species that appear in subsequent ages. Leafy plants come after leafless ones; vertebrates follow invertebrates; brains develop from era to era; and man, the last to come with the exception of his own parasites,[8] is alone among all the animals to have acquired through speech the complete liberty of expressing thoughts, and through fire the power to transform nature.

When we look at the more restricted field of the written history of nations, general progress does not seem so clearly evident. Many defeatists found evidence that humanity does not progress at all, but only shifts, gaining on one side and losing on the other, rising through certain peoples and decaying through others. During the very epoch in which the most optimistic sociologists were preparing the way for the French Revolution in the name of the continuous progress of man, other writers, impressed by the tales of explorers who had been seduced by the simple life of distant peoples, spoke of returning to the mode of existence of these primitives. "Return to nature" was the cry of Jean-Jacques Rousseau. It is strange that this call, however contrary to that of the "Rights of Man and of the Republican," found its way into the language and ideas of the time. The revolutionaries wanted simultaneously to return to the era of Rome and of Sparta, as well as to the happy and pure ages of prehistoric tribes.

In our time, a trend analogous to the "return to nature" movement has emerged, and even more earnestly than in the time of Rousseau. The reason is that current society, which has expanded to the point of including all of humanity, tends to assimilate more intimately the heterogeneous ethnic components from which progressive civilizations remained separated for a long time. Moreover, anthropological studies of the psychology of our primitive brothers have made enormous strides, and the greatest explorers have added to the discussion the decisive weight of their testimony.

We no longer have to rely on such simple and naïve stories as those of Jean de Léry, Claude d'Abbeville, or Yves d'Evreux about the Tupinambá and other Brazilian savages, stories that nevertheless deserve to be greatly appreciated. We also have better statements than the hasty observations of Cook and Bougainville, for the chronicles are now replete with very scrupulous testimonials drawn from long experience. Among the tribes that must undeniably be ranked very highly among men who are closest to the ideal of mutual aid and brotherly love, we must definitely count the Aeta, classified among the primitives, who gave their name "Negros" to one of the Philippine islands.

In spite of all the evils that the whites have done to them, these "Negritos" or "little Negroes" have remained gentle and benevolent toward their persecutors, and it is among them that the virtues of the race are most evident. All members of the tribe think of themselves as brothers, so that when a child is born, the entire extended family gathers to decide on an auspicious name with which to greet the newborn. Their marriages, which are invariably monogamous, depend on the free will of the spouses. The sick, the children, and the elderly are cared for with perfect devotion. No one exerts power, yet all bow willingly to the elderly to show respect for their experience and advanced age.[9] Is there any country in Europe or America that deserves praise equal to this? But we must wonder whether this humble society of the good Aeta still exists. Has it been able to preserve its dwellings of woven branches, its huts of reeds or palms, against the great American hunting party?[10]

Let us take another example from men who have a wider horizon, among populations that are closer to the white race and whose very way of life compels them to pass a large part of their existence away from the maternal hut. The Unangin, referred to by the Russians as the Aleuts after the name of the islands that they inhabit, live in a region of rain, wind, and storms. In order to adapt to their surroundings, they build huts that are half underground, constructed mostly of woven branches covered by a shell of hardened mud and illuminated at the top by a large lens of ice. The necessity of obtaining food has made these Aleuts a fishing people, skilled at maneuvering boats of stretched skins, which they enter as if into a drum. The dangerous seas that they travel have made them intrepid seamen and gifted foreseers of storms. Some of them, especially the whalers, become true naturalists and constitute a special guild whose members are required for initiation to endure a long period of ordeals.[11] The Aleuts, like their neighbors on the mainland, are extraordinarily skillful sculptors, and fascinating objects have been discovered in their burial sites under vaults of rocks. The complexity of Aleut life is also evident in their code of social decorum, which is strictly regulated by custom among blood relatives, relations by marriage, and strangers. Having attained this relatively high degree of civilization, the Aleuts remained, thanks to their isolation, in a state of peace and perfect social equilibrium until a recent period. The first European explorers who made contact with them unanimously praised their good qualities and virtues. Archbishop Innokenti (better known by the name Veniaminov), who witnessed their way of life for ten years, depicted them as "the most affectionate of men" and as beings of incomparable modesty and discretion who are never guilty of the slightest violence in word or deed: "During our years of living together, not one ill-mannered word passed their lips." In this respect, there is certainly no comparison between our people of Western Europe and the little tribe of the Aleutians! The spirit of solidarity and the dignity of moral life among these islanders was so great that some Greek Orthodox missionaries decided not to try to convert them: "What good would it do to teach them our prayers? They are better than we are."[12]

To these examples, chosen from various stages of civilization, can be added equally significant ones from the travels of sociologists and from specialized works in ethnology. Numerous cases can be found in which there is both moral superiority and a more serene appreciation of life among so-called savage or barbarous societies, although these are greatly inferior to ours in the intellectual understanding of things. In the unending spiral that humanity ceaselessly travels, in evolving upon itself in a continuous motion that is roughly comparable to the rotation of the earth, it often happens that certain parts of the larger whole are much closer than others to the ideal focus of the orbit. Perhaps some day the law governing this back-and-forth motion will be understood precisely. For now, it is enough to note the simple facts without drawing premature conclusions and, above all, without accepting the paradoxical views of gloomy sociologists who see in the material progress of humanity only evidence of its actual decline.

Great minds seem at times to have succumbed to this outlook. The following memorable passage from *Malay Archipelago*, published in 1869 by A.R. Wallace, might actually be regarded as a sort of manifesto, a challenge to the ingenuity of those who would unconditionally defend the theory of the continuous progress of humanity. This challenge still awaits a reply. It may be useful to recall his words and to take them as a standard by which to judge historical studies:

> What is this ideally perfect social state towards which mankind ever has been, and still is tending? Our best thinkers maintain, that it is a state of individual freedom and self-government, rendered possible by the equal development and just balance of the intellectual, moral, and physical parts of our nature,—a state in which we shall each be so perfectly fitted for a social existence, by knowing what is right, and at the same time feeling an irresistible impulse to do what we know to be right., that all laws and all punishments shall be unnecessary. . . . Now it is very remarkable, that among people in a very low stage of civilization, we find some approach to such a perfect social state. I have lived with communities of savages in South America and in the East, who have no laws or law courts but the public opinion of the village freely expressed. Each man scrupulously respects the rights of his fellow, and any infraction of those rights rarely or never takes place. In such a community, all are nearly equal. There are none of those wide distinctions, of education and ignorance, wealth and poverty, master and servant, which are the product of our civilization; there is none of that wide-spread division of labor, which, while it increases wealth, products also conflicting interests; there is not that severe competition and struggle for existence. . . . [W]e shall never, as regards the whole community, attain to any real or important superiority over the better class of savages.[13]

But it would be wrong to generalize the observations made by the great naturalist and sociologist about the indigenous peoples of the Amazon and of the Insulindes,[14] and to apply them to all the savage populations of every continent and archipelago. The island of Borneo, where Wallace's view was shaped by so many examples of this moral nobility, is the same great land that Boek has described as the "Land of the Cannibals."[15] One could also call it the "Land of the Headhunters," referring to the men of Dayak who, in order to earn the right to call themselves "men" and to start a family, must chop off one or more heads, whether through trickery or in fair combat. Likewise, the wonderful island of Tahiti, the New Cythera of which eighteenth-century explorers spoke with such naïve enthusiasm, only partly merits the praise of the Europeans who were delighted by both the beauty of the countryside and the friendliness of the inhabitants. Certain august and gentle dignitaries and venerable elders, who in their noble gravity seemed to complete the charming picture of an oceanic paradise, may have belonged to the formidable caste of the Oro (Arioï), which, after having constituted a celibate clergy, became in the end an association of murderers indulging in the infernal rites of killing all their children. It is true that at this point the Tahitians had already reached a level of cultural evolution far beyond the primitive stage. But does this period represent a regression, rather than a development in the direction of progress? Or did the two movements converge in the social life of this little nation locked in its narrow oceanic universe?

Herein lies the main difficulty. Thousands of tribes and other ethnic groupings, lumped together under the name "savages" by haughty "civilized" people, correspond to distinct points that are very different from one another, spaced variously along the path of time and within the infinite network of environments. One tribe is in the middle of a progressive evolution, while the other is obviously in decline. One is in a state of becoming, the other on the road to decay and death. Each of the examples presented by various authors engaged in the general investigation of progress should thus be accompanied by the particular history of the human group in question, for two situations that seem to be almost identical can have an absolutely opposite meaning if the one corresponds to the infancy of an organism and the other to its old age.

One primary fact clearly stands out in comparative ethnographic studies: the essential difference between the civilization of a primitive tribe that is yet only slightly influenced by its neighbors, and the civilization of immense, modern political societies with their unbridled ambition, consists of the simple character of the former and of the complex character of the latter. The first, though not highly developed, at least has the advantage of being coherent and consistent with its ideals. The second is vast, owing to the scope it encompasses, and is infinitely superior to primitive culture in terms of the forces it sets in motion. It is complex and diverse, burdened with survivals from the past, and necessarily incoherent and contradictory. It lacks unity and pursues opposing objectives simultaneously. In prehistoric societies and in those of the world still considered savage, a balance can very easily be established because their ideal is simple.[16] Accordingly, such tribes and primitive races, which have developed very little scientific knowledge, possess only rudimentary crafts and lead a life without much variety; nevertheless, they have been able to attain a level of mutual justice, equitable well-being, and happiness greatly surpassing the corresponding characteristics of our modern societies. The latter are infinitely complex, and are swept along through discoveries and partial progressions in a continual momentum of renewal that blends in various ways with all of the factors from the past. Also, when we compare our powerful, global society to the small, almost unnoticeable groups of primitives who have managed to maintain themselves apart from the "civilizers"—who are all too often destroyers—we might be led to conclude that these primitives are superior to us and that we have regressed over the course of time. But our acquired qualities are not of the same order as the ancient ones, so it is very difficult to make an equitable comparison. Society has greatly increased its baggage since primitive times. In any case, it is very agreeable to focus on the dozens or hundreds of individuals who have developed harmoniously within the limits of their narrow cosmos, and who were fortunate enough to realize on a small scale that which we are now trying to accomplish at the level of the entire human universe. In societies in which all know each other as members of the same family, the desired goal is near at hand. It is different for our modern society, which encompasses a world but does not yet embrace it.

If we look at humanity in its entirety, and even return to the origins of living beings, we can regard all social groupings as normally forming small, distinct colonies, from the floating ribbons of salpa on the sea, to the swarms of bees that gather at the same hive, to peoples who seek to demarcate themselves precisely within borders. The earliest groupings are microcosmic, and then they become more and more extended and increasingly complex over time, to the degree that an ideal arises and becomes more difficult to achieve. Each of these small societies constitutes by nature an independent and self-sufficient organism. However, none of them are completely closed, except for those that are isolated on islands, peninsulas, or in mountain cirques whose access has been cut off. As groups of men encounter one another, direct and indirect relations arise. In this way, following internal changes and external events, each swarm ends its particular, individual evolution and joins willingly or forcibly with another body politic so that both are integrated into a superior organization with a new course of life and of progress before it. This metamorphosis is analogous to that by which a seed changes into a tree, or an egg into an animal: there is a transformation from homogeneous to heterogeneous structure.[17] But diverse outcomes are possible. Among small, isolated societies, a great number perish from senile exhaustion through a bloody conflict before realizing the more or less exalted end toward which their normal functioning tends. Other microcosms, having an environment more conducive to their harmonious development, are able to attain their ideal successfully and live according to the rules of wisdom established by their ancestors. Thus a number of tribes that had a simple social organization and a naïve general conception of the universe, and that were free from mixture with other ethnic components, succeeded in constituting small cells of perfected form and well-arranged organs. Each individual was conscious of his solidarity with all the other members of the tribe and enjoyed through each individual an absolutely respected personal liberty, an inviolate justice, and a calm and tranquil life. These tribes have come close to the state that one could call "happiness" if this word were to imply only the satisfaction of instincts, appetites, and feelings of affection.

In the history of humanity, several social types have successively reached their full blossoming. Similarly, among the more ancient worlds of flora and fauna, numerous genera and species have reached such ideals of strength, rhythm, or beauty that nothing superior to them can be imagined. While the rose is the precursor of many subsequent forms, it is no less perfect or insurpassable for it. And among animals, is it possible to imagine any organisms more definitive, each of their kind, than crinoids, beetles, swallows, antelopes, bees, and ants?[18] Is man, still imperfect in his own eyes, not surrounded by countless living beings that he can admire unreservedly if he has open eyes and an open mind? And even if he chooses among the infinite number of types around him, does he not in reality do so through his inability to embrace everything? For each form, epitomizing in itself all of the laws of the universe that converge to determine it, is an equally marvelous consequence of this process.

Therefore, modern society can lay claim to a particular superiority over the societies that preceded it only through the greater complexity of the elements that enter into its formation. It has a greater scope and constitutes a more heterogeneous organism through the successive assimilation of juxtaposed organisms. But on the other hand, this vast society tends to become more simplified. It seeks to realize human unity by gradually becoming the repository of everything achieved from labor and thought in all countries and all ages. Whereas the various tribes living separately represent diversity, the nation whose aim is preeminence over and even the absorption of other ethnic groups tends to achieve great unity. In effect, it seeks to benefit by the resolution of all conflicts, and to create one unified truth out of all the small, scattered truths. But the road that leads to this goal is very difficult, full of obstacles, and, above all, criss-crossed with deceptive paths that seem at first to be parallel to the main route that we fearlessly take! History has shown us how each nation, no matter how well endowed, strong, and healthy it may be in its prime, ends up lagging behind after a number of decades or centuries and then disintegrates into smaller bands that wander off, scattering across the surrounding countryside. Sometimes it even tries to return to its origins, but the diversity of languages, of factions, and of local interests prevails over the feeling of human unity, which for a time sustains the nation in its progress.

In our time, the idea of human unity has so deeply penetrated various civilized ethnic groups that they are, so to speak, immunized against decline and death. Barring great cosmic revolutions whose shadows have yet to fall over us, modern nations will in the future escape the phenomena of seemingly final ruin that occurred to so many ancient peoples. Certainly, political "transgressions," analogous to marine transgressions on coastlines, will occur on the borders of states, and these borders themselves will disappear in many places, prefiguring the day when they will cease to exist everywhere. Various geographical names will be erased from maps, but despite such changes, the peoples encompassed by modern civilization (which covers a very considerable portion of the earth's land surface) will certainly continue to participate in the material, intellectual, and moral progress of one another. They are in the era of mutual aid, and even when they engage in bloody conflicts with each other, they do not stop working in part for the common welfare. During the last great European war between France and Germany, hundreds of thousands of men perished, crops were devastated, and wealth was destroyed. Each side despised and damned the other, but that did not in the least prevent either side from continuing the labor of thought for the benefit of all men, including mutual enemies. There were patriotic disputes over whether the diphtheria serum had been effectively discovered and applied for the first time to the east or west of the Vosges, but in France as in Germany, the medicine increased the power of a unified humanity over an indifferent nature. In a similar way, a thousand other new inventions have become the common heritage of the two neighboring nations—rivals and enemies, it is true, but still fundamentally very close friends since they engage relentlessly in broader work for the benefit of all men. And in the Far East, one finds that the covert or overt war between Japan and Russia cannot stop the astonishing progress that is being accomplished in this part of the world through the sharing of human culture and ideals. A historical period has already earned the name of "humanism" because at that time the study of Greek and Latin classics united all refined men in the common appreciation of great thoughts expressed in fine language. Our epoch is even more deserving of such a name since today it is not only a brotherhood of intellectuals who are joined together but also entire nations descended from the most diverse races and peopling the most distant parts of the world!

Yet in our time, a fatuous humanitarianism [*humanitairerie*] is quite prevalent. All statesmen and great writers make fun of this poor sentimentality. The second half of the nineteenth century was fertile in theories about the forms progress sometimes takes. For example, the revolutionaries of 1848 proclaimed with extraordinary brilliance the idea of "humanity." But in their profound ignorance, these brave souls had no idea of the difficulties that their propaganda would have to encounter, and, moreover, it was easy after their defeat to ridicule them. Then came the Franco-Prussian War, the crowning glory of Bismarckian politics, which came to fruition in a sentimental Germany. Everyone vied with one another to imitate, with equal ineptitude, the machinations of the Iron Chancellor, whose shadow still looms over us. The liberation of Greece and the Two Sicilies,[19] and the acclaim that greeted Byron, Kossuth, Garibaldi, and Herzen was followed by the most restrained conduct in response to the massacres in Armenia, the slaughter in eastern Africa, and the pogroms of Russia.[20] A passionate nationalism rages in all western countries, and existing borders have for the most part been tightened during the past fifty years. We have also seen in Great Britain the republican idea, which united many supporters before 1870, gradually fade from the political scene. It is the same in all civilized countries for the most idealistic of "utopias." One can thus become discouraged by classifying these distinct evolutions as definite regressions if one does not also investigate their causes. Once it is understood how this movement of reversal functions, there can be no doubt that the cry of humanity will once again resound when the "weak and the downtrodden" (who have never stopped proclaiming this ideal among themselves) will have acquired a thorough scientific knowledge. Having attained a more complete mastery of international understanding, they will feel strong enough to abolish forever all threat of war.

Conflicts between rival governments can be serious and full of repercussions; however, even when these disputes lead to war, they cannot have results analogous to those of the struggles that long ago destroyed the Hittites, the Elamites, the Sumerians and Akkadians, the Assyrians, the Persians, and before them so many civilizations whose very names are unknown to us. In reality, all nations, including those that call themselves enemies, and in spite of their leaders and the survival of hatreds, form but one single nation in which all local progress reacts upon the whole, thus contributing to general progress. Those whom the "unknown philosopher" of the eighteenth century called "men of desire"—in other words, men who desire good and who work toward its realization—are already sufficiently numerous, active, and harmoniously grouped into one moral nation for their labor of progress to prevail over the elements of regression and separation produced by surviving hatreds.

It is this new nation, composed of free individuals, independent from one another but nonetheless amicable and unified, that must be addressed. It is to this humanity in formation that we must direct propaganda on behalf of all the reforms that are desired and all the ideas that seem just and renewing. This great nation has expanded to all corners of the earth, and it is because it is already aware of itself that it feels the need for a common language. It is not acceptable that these new fellow citizens should merely speculate about one another from one end of the earth to the other—they must understand each other completely. We can be confident that the language that we hope for will come into being: every strongly willed ideal can be realized.

This spontaneous union across borders of men of good will removes all authority from certain falsely named "laws" that were generalized from previous historical evolution and that now deserve to be relegated to the past as having had only relative truth. One example is the theory according to which civilization was supposed to have made its way around the earth from east to west, like the sun, and determined its focus from millennium to millennium on the circumference of the planet. Some historians, struck by the elegant parabola traced by the spread of civilization between ancient Babylon and our modern Babylons, formulated this law of the precession of culture; however, before the flowering of Hellenic culture, the Egyptians, in seeking to comprehend the vastness of their Nilotic world, a true universe unto itself by virtue of its extent and its isolation, attributed a quite different direction to the propagation of human thought. They believed that it had come to them from south to north, carried like fertile alluvial soils by the waters of the Nile. They were probably wrong, and in at least one known historical epoch, civilization spread in the opposite direction, from Memphis toward Thebes with its "Hundred Doors."[21] In other lands, the movement of culture proceeds downstream along rivers and successively gives rise to populous cities that are centers of human labor. Similarly, in India the trajectory is from northwest to southeast along the banks of the Ganges and the Jamuna, and on the vast plains of China, the "line of life" clearly travels from east to west through the valleys of the Huang He and the Chang Jiang.

These examples suffice to show that the so-called law of progress determining the successive transfer of the predominant global focus of progress from east to west has only a provisional and localized validity, and that other serial movements have prevailed in various regions, depending on the slope of the terrain and the forces of attraction produced by environmental conditions.[22] Nevertheless, it is good to recall the classic thesis, not only in order to understand the causes that gave rise to it, but also because it is still invoked by an ambitious nation of the "Great West," which loudly proclaims its right to preeminence.[23] But has it not become obvious to the members of the great human family that the center of civilization is already everywhere, by virtue of a thousand discoveries and their applications that occur every day in one place or another and then spread immediately from city to city across the surface of the earth? The imaginary lines that history once traced over the globe have been submerged, so to speak, by the waves of the deluge that now covers all countries. This deluge is really the flood of knowledge that the gospel says (albeit from a different point of view) ought to spread equally over all parts of the earth. The element of distance has lost its importance, for man can and indeed does educate himself about all the phenomena relating to soil, climate, history, and society that distinguish different countries. Now to understand one another is to be already associated, to be intermingled to a certain extent. Certainly, there are still contrasts between different lands and different nations, but these contrasts are diminishing and tend gradually to be neutralized in the minds of the well-informed. The focus of civilization is wherever one thinks or acts. It is in the laboratory in Japan, Germany, or America where the properties of a particular metal or chemical substance are discovered, in the plant where propellers for ships or aircraft are built, or in the observatory where previously unknown data concerning the movement of the stars are recorded.

The once-famous theory of Vico on the *corsi* and *ricorsi* (ebb and flow) of historical evolution is now as much out of favor as the theory of the successive displacement of centers of culture. A closed society behaving like a single individual would no doubt have a natural tendency to develop according to rhythmic oscillations, with periods of activity following periods of rest, and, whenever the process would resume, the action of the same elements under similar conditions would bring about an almost identical operation. The alternation from democracy to a tyrannical regime and from tyranny back to popular government would thus occur with a swinging motion similar to that of a clock's pendulum. But as our knowledge of history grows, and as ethnic factors become more influential in various ways, we see that such rhythmic alternation of events is inevitably disturbed: the ebb and flow take on such amplitude and merge in such a varying manner that they cannot clearly be distinguished. It was largely to establish the proper relationship between them that the two-dimensional model of Vico's swinging pendulum was replaced by an infinite curve ascending in spirals. Here is just the sort of poetic image that Goethe was fond of sketching; however, it corresponds only vaguely to reality. It is true that when the infinite entanglement of historical facts is studied from a distance, they seem to form themselves into large masses. But beneath the surface there is a constant movement of action and reaction, and the sum of the various conflicting forces can never carry humanity along a straight line. The whole of this vast profusion certainly does not lack harmonious development, and there are remarkable regularities in the thousand changing details of its scenes. But however elegant geometrical forms may seem, they cannot give an adequate idea of its endless undulations.

The extension of the scope of research, which increases through revolutions and the passage of time, constitutes one of the principal elements of progress. Self-conscious humanity has grown continuously in proportion to the geographical assimilation of distant lands into the realm of those already scientifically examined. Whereas the explorer conquers space, thus allowing men of good will to unite their efforts throughout the world, the historian, turning toward the past, conquers time. Humankind, which makes itself One at every latitude and longitude, similarly tries to realize itself through one form that encompasses all ages. This is a conquest no less important than the first. All past civilizations, even those of prehistory, offer us a glimpse of the treasure of their secrets and, in a certain sense, are gradually merging into the life of present-day societies. We can now look back on the succession of epochs as one synoptic scene that plays out according to an order in which we can seek to discover the logic of events. In doing so, we cease to live solely in the fleeting moment, and instead embrace the whole series of past ages recorded in the annals of history and discovered by archeologists. In this way, we manage to free ourselves from the strict line of development determined by the environment that we inhabit and by the specific lineage of our race. Before us lies the infinite network of parallel, diverging, and intersecting roads that other segments of humanity have followed. And throughout this series of epochs stretching out toward an indefinite horizon, we find examples that appeal to our spirit of imitation. Everywhere we see brothers toward whom we feel a growing spirit of solidarity. As our overview of history extends ever further into the past, we find an increasing number of models demanding understanding, including many that awaken in us the ambition to imitate some aspect of their ideal. As humanity became more mobile and modified itself in the most diverse ways, it lost a significant part of its achievements attained in the past. Today, we may ask whether it is possible to recover all of the baggage we have left at the various stations of our long voyage through the centuries.

Since men are henceforth masters of time and space, they see an infinite field of achievement and progress opening before them. However, burdened by the illogical and contradictory conditions of their surroundings, they are hardly in a position to proceed knowledgeably with the harmonious work of improvement for all. This is understandable. All initiative comes from individuals and insignificant minorities, and these isolated persons or small groups attend to the most urgent needs first, directly attacking whatever evil they find before them. So if their efforts have the advantage of emerging simultaneously on almost all fronts, by the same token, they lack coherent strategy. But theoretically, when one detaches oneself intellectually from the chaos of conflicting interests, it is easy to see immediately that the true and fundamental conquest, from which all others can logically be derived, is that of procuring bread for all men—for all who call themselves "brothers," even though they are very far from being so. When all have enough to eat, all will feel that they are equal. Now this is precisely the ideal that many a small tribe far from our great pathways of civilization already knew how to realize, and we must come to terms with this ideal of solidarity as soon as possible if all of our hopes for progress are not to become the most cruel of ironies. Montaigne has described the opinion on this subject held by the Brazilian natives who were brought to Rouen in 1557 "at the time that the late King Charles the Ninth was there."[24] They were struck by many strange things and above all by the fact "that there were among us men full and crammed with all sorts of good things, [for] which their halves [fellow countrymen] were begging at their doors, emaciated with hunger and poverty; and they thought it strange that these necessitous halves were able to suffer such an injustice, and that they did not take the other by the throat or set fire to their houses."[25] For his part, Montaigne greatly pitied these savages from Brazil for "allowing themselves [to] be deluded with desire of novelty and to leave the serenity of their sky to come and gaze at ours!"[26] They were "unaware . . . that from this intercourse will be born their ruin."[27] Indeed, these Tupinambá from the American coast have left not a single descendent. All of the tribes were exterminated, and if there still remains a little blood of these indigenous people, it is mixed with that of some despised proletarians.

The conquest of bread, which true progress requires, must be an actual conquest.[28] It is not simply a question of eating, but of eating the bread that is due by human right rather than owing to the charity of a great lord or wealthy monastery. The unfortunate people who beg at the doors of the barracks and churches number in the hundreds of thousands, perhaps in the millions. Thanks to the vouchers for bread and soup distributed by charity, they barely manage to get by; however, it is very unlikely that the aid provided for all these needy people has had the slightest significance in the history of civilization. The very fact that they have been fed without having asserted their right to food, and perhaps even required to express their gratitude, proves that they consider themselves to be simply the dregs of society. Free men look each other in the eye, and the first condition of their forthright equality is that individuals be absolutely independent of one another, and that they earn their bread through a mutuality of services. Entire populations have been reduced to moral ruin through a gratuitous material existence. When Roman citizens lived in a state of abundance and did not have to work for the food and entertainment provided by the masters of the state, did they not stop defending the empire? A number of classes, among them that of the "deserving poor," prove completely useless in relation to progress as a result of the system of alms, and some cities have fallen into irreversible decay because they contain an idle multitude that, having no need to work for itself, also refuses to work for others. This is the real reason that so many cities and even nations are "dead." Charity brings with it a curse on those it nourishes. This can be witnessed in the Christmas celebrations of the aristocracy, in which young heirs to vast fortunes, draped in luxurious clothes, practice their noble gestures and gracious smiles. And then, under the loving eyes of their mothers and governesses, they nobly distribute presents to the poor of the streets, who are dutifully washed and dressed in their Sunday best for the occasion. Is there a spectacle sadder than that of these young unfortunates, stupefied by the glory of gold in all its munificence?

Down with this ugly Christian charity! The cause of progress is entrusted to the conquerors of bread—in other words, to the working people who are united, free, equal, and released from the bonds of patronage. It will be up to them to finally use scientific method in applying each discovery to the interests of society, and to realize Condorcet's assertion that "Nature has placed no limit on our hopes." For, as another historian and sociologist said, "The more one asks of human nature, the more it gives. Its faculties are stimulated by effort, and its power seems unlimited."[29] As soon as man is firmly confident of the principles according to which he directs his actions, life becomes easy. Fully aware of his due, he accordingly recognizes that of his neighbor. In doing so, he brushes aside the functions usurped by the legislature, the police, and the executioner; thanks to his own ethic, he abolishes law (Emile Acollas). Self-conscious progress is not a normal function of society, a process of growth analogous to that of a plant or animal. It does not open like a flower;[30] instead, it must be understood as a collective act of social will that attains consciousness of the unified interests of humanity and satisfies them successively and methodically. And this will becomes ever stronger as it surrounds itself with new achievements. Once accepted by all, certain ideas become indisputable.

The essence of human progress consists of the discovery of the totality of interests and wills common to all peoples; it is identical to solidarity. First of all, it is necessary to address the economy, which is very different from that of primitive nature, in which the seeds of life pour out with astonishing abundance. At present, society is still very far from achieving the wise use of forces, especially human forces. It is true that violent death is no longer the rule as in former times. Nevertheless, the vast majority of people die before their time. Disease, accidents, injuries, and defects of all kinds, most often complicated by medical treatments applied wrongly or randomly and exacerbated above all by poverty, the lack of essential care, and the absence of hope and cheer, cause decrepitude long before the normal onset of old age. Indeed, an eminent physiologist[31] has written a wonderful book whose principal thesis is that almost all old people die before their time and with an absolute dread of death, which would instead arrive like sleep if it were to come at a time when a man, happy to have led a good life full of activity and love, felt the need for rest.

This uneconomical use of forces is demonstrated above all in great changes, such as violent revolutions and the introduction of new processes. Old equipment, as well as men who are accustomed to a previous form of labor, are discarded as useless; however, the ideal is to know how to utilize everything, to employ refuse, waste, and slag, for everything is useful in the hands of one who knows how to work with the materials. Generally speaking, all modification, no matter how important, is accomplished through a combination of progress and a corresponding regression. A new organism is established at the expense of the old. Even when the vicissitudes of conflict are not followed by destruction and ruin in the strictest sense, they are nevertheless a cause of local decline. The prosperity of some brings the downfall of others, thus confirming the ancient allegory that depicts Fortune as a wheel, lifting up some while crushing others. The same fact can be evaluated in many ways: on the one hand as a great moral advance, and on the other as evidence of decay. From a great, fundamental event such as the abolition of slavery, disastrous consequences can ensue due to the thousand blows and counterblows of life, contrasting with the totality of fortunate results. The slave, and generally speaking even the man whose life has been regulated from infancy and who has never learned to distinguish clearly between two successive and very distinct states of his milieu, easily becomes accustomed to the unchanging routine of existence, as mundane as it may be. He can live without complaining, like a stone, or like a plant hibernating under the snow. As a result of this habituation, during which thought slumbers, it often happens that the man who is suddenly liberated from some form of servitude does not know how to accommodate himself to his new situation. Not having learned how to exercise his will, he stares like an ox at the stick that once goaded him to work. He awaits the bread that had always been thrown to him and that he was accustomed to picking up from the mud. The qualities of slavery, obedience and resignation—as far as one can call them "qualities"—are not the same as those of the free man: initiative, courage, and indomitable perseverance. The person who retains even vaguely the first qualities, who allows himself to miss his former life ruled by the carrot and the stick, will never be the proud hero of his destiny.

On the other hand, the man who has cheerfully accommodated himself to the conditions of a new life of perfect independence, a life that gives to the agent full responsibility for his conduct, is in danger of unimaginable suffering when he finds himself caught again in a vestige of ancient slavery—the military, for example. His life then becomes unbearable, and suicide seems like a refuge. Thus in our incoherent society, in which two opposing principles struggle against one another, it is possible to desire death either because it is too difficult to conquer life or because liberty has so many joys that one cannot give them up. Is it not contradictory that the reaction to a greater intensity in life can be an extraordinary increase in bouts of despair and an obsessive fear of death? The number of suicides has continually increased for several decades in contemporary society and in all so-called civilized countries. Not long ago, this type of death was rare in all lands and completely unknown among certain peoples such as the Greeks, for whom, moreover, poverty, temperance, and harsh work were the rule. But the great whirlwind generated by the cities has produced a corresponding torrent of passions, emotions, changing impressions, ambitions, and insanity in our modern "Babylons." Since life is more active and passionate, it is frequently complicated with crises and often ends abruptly through voluntary death.

This is the very sorrowful aspect of our much-acclaimed half-civilization (it is only half-civilized because it is far from benefiting everyone). The average man of our time is not only more active and lively but also happier than in previous times when humanity, divided into innumerable tribes, had not yet become conscious of itself as a whole; however, it is no less true that the moral discrepancy between the way of life of the privileged and that of the outcasts has increased. The unfortunate have become more unfortunate, and envy and hatred are added to their poverty, increasing their physical suffering and forced deprivation. In primitive clans, the victims of starvation and sickness are subject only to physical pain. But among our civilized people, they must also bear the burden of humiliation and even public loathing. Their living conditions and clothing make them seem sordid and repugnant to the observer. Are there not neighborhoods in every large city that are carefully avoided by travelers because of an aversion to the nauseating odors that emanate from them? Except for the Eskimos in their winter igloo, no savage tribe inhabits such hovels as exist in Glasgow, Dundee, Rouen, Lille, and so many other industrial cities, where in cellars with slimy walls, beings that resemble humans drag themselves about painfully for a time in a semblance of life. The barbaric Hindus who live in the forests at the center of the subcontinent, clothed in a few colorful rags, offer a relatively cheerful sight compared to these emaciated proletarians of luxurious Europe, somber, sad, and gloomy in their tattered, filthy clothes. For the observer who is not afraid to go near the factories when they let out, the most striking thing, aside from the clothing of poverty, is the absolute absence of personality. All these beings rushing toward an inadequate meal have had since youth the same withered face and the same vacant, deadened stare. It is impossible to distinguish among them any more clearly than among sheep in a flock. They are not humans, but rather arms, or "hands," as they are so appropriately called in the English language.

This horrible discrepancy, this most dreadful scourge of contemporary society, could be corrected rapidly by scientific method through the redistribution of the goods of the earth, since the resources necessary for all humans are in superabundance. This goes without saying. Humanity is admirably equipped through its progress in the knowledge of time and space, of the innermost nature of things, and of man himself. But is it currently advanced enough to tackle the fundamental problem of its existence, which is the problem of the realization of its collective ideal, not only for the "ruling classes," one caste, or a group of castes, but for all whom a religion once described as "brothers created in the image of God"? Of course, humanity can reach this goal. There will no longer be a question of hunger the day that people who are starving join together to claim their due.

Similarly, the question of education will be resolved, since the problem is acknowledged in principle and because the desire for knowledge is widespread, even if it is only in the form of curiosity. Now one advancement never comes alone; it has a complementary and reciprocal relationship with other advancements in the entirety of social evolution. As soon as the sense of justice is satisfied through the participation of all in the material and intellectual resources of humanity, each man will as a result experience a great unburdening of his conscience. For the present cruel state of inequality, in which some are overloaded with superfluous wealth while others are deprived even of hope, weighs like a bad conscience on the human soul, whether one is aware of it or not. It weighs most on the souls of the fortunate, whose joys are always poisoned by it. The greatest step toward peace would be for no one to do wrong to his neighbor, for it is in our nature to hate those whom we have wronged and to love those whose presence recalls our own worth. The moral consequences of the very simple act of justice in which bread and education are guaranteed to all would be incalculable.

If, continuing the present direction of historical evolution, humanity soon reaches the goals of abolishing death from hunger and stagnation from ignorance, then another ideal will appear like a shining beacon—an ideal that moreover is already being pursued by an ever-growing number of individuals. This is the lofty ambition to regain all lost energies, to prevent the loss of present forces and materials, and also to recover from the past everything that our ancestors allowed to slip away. Generally speaking, this would mean that civilizations would imitate the engineers of our day who are discovering treasures in the debris that was considered worthless by the Athenian miners of the past. If it is true that in certain respects some primitives and ancients surpassed the average modern-day man in strength, agility, health, and beauty, then we must become their equals! Granted, this reconquest will not go so far as the recovery of the use of atrophied organs whose former purposes have been discovered by biologists (such as Elie Metchnikoff); however, it is important to know how to maintain fully those energies that are still accorded to us and to retain the use of muscles that, while continuing to function, have become less flexible and are in danger of soon becoming worthless to our bodies. Is it possible to prevent this physical diminishment of man, who is thrown out of balance by the development of his mental capacities? It is predicted that man will gradually turn into an enormous brain, wrapped in bandages to protect him from colds, and that the rest of his body will atrophy. Is there anything we can do to resist this tendency? Zoologists tell us that man used to be a climbing animal, like the monkey. Why, then, does modern man let himself forfeit this skill of climbing, which certain primitives still possess to a remarkable degree, notably those who climb to the tops of palm trees to gather bunches of fruit? As mothers never fail to observe admiringly, infants have astonishing grasping power, with which they can suspend their bodies, even for minutes at a time,[32] yet they gradually lose this initial strength because great care is taken to deny them the opportunity to exercise it. The threat of clothing being ripped and torn through the child's efforts to climb are enough for the parents of our economically-minded society to forbid their offspring to climb trees. The fear of danger is only a secondary consideration in this prohibition.

As a result of such fears, most "civilized" children remain greatly inferior to the sons of savages in games of strength and agility. Furthermore, since they have had little opportunity to exercise their senses outdoors, they do not have the same clarity of vision or keenness of hearing. Compared to the animals of beautiful form and sharpened senses that Herbert Spencer thought they should be, they seem for the most part to have clearly degenerated. In no way do they merit the words of admiration evoked in European travelers by the sight of the young men of Tenimber, practicing stringing their bows or throwing the javelin.[33] The players of pelote, golf, and lacrosse constitute the elite of civilized people for physical beauty. But the spectators would have difficulty finding perfectly balanced forms to rhapsodize over, even among the champions. The evidence is clear. It is certain that in purity of line, dignity of bearing, and gracefulness of movement, a number of Negro, American Indian, Malayan, and Polynesian tribes surpass randomly selected groups representing the average type of the nations of Europe, though perhaps not certain exceptional cases among Europeans. Thus, from this perspective there has been a general regression because of our confinement to our homes and our absurd clothing, which interferes with perspiration, the effect of air and light on the skin, and the free development of muscles, which are often constricted, tortured, or even crippled by laced boots and corsets. Nevertheless, numerous examples prove that this regression is not final and irrevocable, since our young people who have been raised in good hygienic conditions and who engage in physical exercise develop in shape and strength like the most beautiful of savages. Besides, they have been granted the superiority of self-awareness and the distinction of intellect. Thanks to the achievements of the past, which moderns acquire rapidly and methodically through education, they succeed in living longer than the savage since they know how to compress into their lives a thousand prior existences and to recall survivals from the past in order to make a logical and beautiful whole out of current practices and the innovations of previous times. If only we could gauge the degree of strength that the modern can attain by using as an example today's skilled mountain climbers of the Alps, the Caucasus, the Rocky Mountains, the Andes, the Tien Shan, and the Himalayas! Certainly, a Jacques Balmat would not have climbed Mont Blanc if a de Saussure had not existed to train him in this undertaking. Today, such experts as Whymper, Freshfield, and Conway are in strength, endurance, knowledge, and the practice of mountain climbing the equals and even the superiors of the most dependable mountain guides, who were trained from youth in all the physical and moral qualities necessary for dangerous ascents. It is the man of science who is now followed by the native to the summit of Kilimanjaro or of the Aconcagua, and it is he who leads the Eskimos to the conquest of the North Pole. Thus it is possible for modern man to realize perfectly his imagined ideal, that of being able to acquire new qualities without losing, or even while regaining, those possessed by his ancestors. This is not at all a chimera.

This strength of understanding, this increased capacity of modern man, permits him to reconquer the past from the savage in his natural, ancient environment, and then to unite it and blend it harmoniously with his own more refined ideas. But all of this increase in strength will result in a permanent, well-established reconquest only on the condition that the new man include all other men, his brothers, in the same feeling of unity with all things.

Here, then, is the social question that is posed anew in its full scope. It is impossible to love wholeheartedly the primitive savage in his natural environment of forests and streams if at the same time one does not love the men living in the more or less artificial society of the contemporary world. How can we admire and love the small, charming individuality of the flower, or feel brother to the animals and approach them as St. Francis of Assisi did if we do not also see our fellow men as beloved companions? The alternative is to avoid them in the name of love so as to escape the moral wounds inflicted by the hateful, the hypocrites, or the indifferent. The complete union of the civilized with the savage and with nature can take place only through the destruction of the boundaries between castes, as well as between peoples. Each individual must be able to address any of his peers in complete brotherhood, and to speak freely with them "about all that is human," as Terence said, without succumbing to the customs and conventions of the past. Life, restored to its original simplicity, thus entails a complete and amicable freedom of human social intercourse.

Has humanity made any real progress along this road? It would be absurd to deny it. What is called the "tide of democracy" is nothing other than the growing feeling of equality among the members of different castes that were recently enemies. Under a thousand changing surface appearances, the work is carried out in the depths, in all nations, thanks to man's growing knowledge of himself and of others. Increasingly he succeeds in finding the common basis for our likeness to one another and manages to extricate himself from the entanglement of superficial opinions that have kept us separated. We march, then, toward future conciliation, toward a form of happiness far more ample than that which satisfied our ancestors, the animals and the primitives. Our physical and moral world has grown larger at the same time that our conception of happiness has become broader. Indeed, in the future, happiness will be considered as such only if it is shared by all, if it is made conscious and is well thought out, and if it includes within itself the fascinating pursuits of science and the joys of antique beauty.

All of this removes us noticeably from the theory of the "Superman" as understood by the aristocrats of thought. The kings and the powerful readily imagine that there are two systems of morals—theirs, which consists of capriciousness; and obedience, which is suitable for the masses. Similarly, arrogant young people who worship the intellectual powers they think they possess, indulgently place themselves on a high terrace of the ivory tower, beyond the reach of humble mortals. They condescend to chat only with a select few. Perhaps they even believe themselves to be alone. Genius weighs heavily upon them. Underneath their inevitably furrowed brows, a turbulent world rages. They are oblivious to the teeming, formless mass of the unknown multitude far beneath the flight of their thought. It is true that man can discover no limits that he cannot surpass through his striving to study and learn. Yes, he must try to realize his own ideal, to seek to surpass it, and to climb ever higher. Even as a dying man, I believe in my personal progress; those who feel as if they are moribund might as well die. But in order to surpass his limits, man does not need to break the bonds that connect him with the beings around him, for he cannot escape the close solidarity that supports his life through the lives of his fellow creatures. To the contrary, each of his personal advancements means progress for those around him: he shares his knowledge as he shares his bread, and he does not leave behind the poor and the crippled. He has had teachers—since he was hardly born without a father like some god in a fable—and he will in turn teach those who come after him.

The barbarous methods of the Spartans are still favored by those ineffectual persons who know neither how to heal nor how to teach. They smother those who seem weak and throw the malformed into a hole, breaking their bones. Such are the summary practices of the ineffectual and the ignorant. And what doctor, midwife, or infallible arbitrator will tell us which newborn can be spared and which is beyond hope? Often, the science practiced by these judges has been faulty. A particular body that they had deemed ill-suited for life actually turned out to be admirably adapted to it. A particular intelligence that from the heights of their judicial bench they had classified as moronic developed brilliant and creative powers. Being old, slaves to routine, and misoneistes,[34] they were completely wrong, and it is through revolution against them that the world was ennobled and renewed. The best approach is to accept all men as equals in potential and in dignity, to help the weak by supporting them with one's own strength, to help restore health to the sick, and to open the minds of the unintelligent to elevated thoughts, all with constant concern for the betterment of others and of oneself. For we are part of a whole, and evolution takes place throughout the world, whether it moves from progress to progress or from regression to regression.

Thus happiness, as we understand it, does not consist simply of personal enjoyment. Of course it is individual in the sense that "each is the artisan of his own happiness," but it is true, deep, and complete only when it extends to the whole of humanity. It is not possible to avoid sorrow, accidents, sickness, or even death; however, by joining together with others in an undertaking whose significance he grasps, and by following a method that he knows to be effective, man can be certain of directing the whole great human body toward the greatest good. In comparison to this body, each individual cell is infinitely small, a millionth of a millionth, counting the present population of the earth and all previous generations. Happiness does not mean the attainment of a certain level of personal or collective existence. It is rather the consciousness of marching toward a well-defined goal to which one aspires and that one creates in part through one's own will. To develop the continents, the seas, and the atmosphere that surrounds us; to "cultivate our garden" on earth; to rearrange and regulate the environment in order to promote each individual plant, animal, and human life; to become fully conscious of our human solidarity, forming one body with the planet itself; and to take a sweeping view of our origins, our present, our immediate goal, and our distant ideal—this is what progress means.

Thus we can with complete confidence respond to the question that arises in the depths of each man's being: yes, we have progressed since the time when our ancestors left their maternal caves, during the several thousand years that make up the brief self-conscious period of human life.

18

Advice to My Anarchist Comrades (1901)

Reclus wrote the following letter on the occasion of the opening of an anarchist congress. It was subsequently published in *Il Pensiero* (June 16, 1907), in *Réveil de Genève* (January 7, 1911), and in volume 3 of *Correspondance*, 238–40.

To the Editors of *la Huelga General* in Barcelona
Brussels, December 4, 1901

Dear comrades,

It is our usual habit to exaggerate both our strengths and our weaknesses. During revolutionary periods, it seems that the least of our actions has incalculably great consequences. On the other hand, during times of stagnation, even though we have dedicated ourselves completely to the cause, our lives seem barren and useless. We may even feel swept away by the winds of reaction.

What then should we do to maintain our intellectual vigor, our moral energy, and our faith in the good fight?

You come to me hoping to draw on my long experience of people and things. So as an old man I give you the following advice.

Do not quarrel or deal in personalities. Listen to opposing arguments after you have presented your own. Learn how to remain silent and reflect. Do not try to get the better in an argument at the expense of your own sincerity.

Study with discretion and perseverance. Great enthusiasm and dedication to the point of risking one's life are not the only ways of serving a cause. It is easier to sacrifice one's life than to make one's whole life an education for others. The conscious revolutionary is not only a person of feeling, but also one of reason, for whom every effort to promote justice and solidarity rests on precise knowledge and on a comprehensive understanding of history, sociology, and biology. Such a person can incorporate his personal ideas into the larger context of the human sciences, and can brave the struggle, sustained by the immense power he gains through his broad knowledge.

Avoid specialization. Side neither with nations nor with parties. Be neither Russians, Poles nor Slavs. Rather, be men who hunger for truth, free from any thoughts of particular interests, and from speculative ideas concerning the Chinese, Africans, or Europeans. The patriot always ends up hating the foreigner, and loses the sense of justice that once kindled his enthusiasm.

Away with all bosses, leaders, and those apostles of language who turn words into Sacred Scripture. Avoid idolatry and value the words even of your closest friend or the wisest professor only for the truth that you find in them. If, having listened, you have some doubts, turn inward toward your own mind and reexamine the matter before making a final judgment.

So you should reject every authority, but also commit yourself to a deep respect for all sincere convictions. Live your own life, but also allow others the complete freedom to live theirs.

If you throw yourself into the fray to sacrifice yourself defending the humiliated and downtrodden, that is a very good thing, my companions. Face death nobly. If you prefer to take on slow and patient work on behalf of a better future, that is an even better thing. Make it the goal of every instant of a generous life. But if you choose to remain poor among the poor, in complete solidarity with those who suffer, may your life shine forth as a beneficent light, a perfect example, a fruitful lesson for all!

Greetings, comrades.
Elisée Reclus

Notes

Preface to the PM Press Edition

1 Perhaps by chance, Nature's gaze seems to be focused in the direction of New Orleans, the location at which the translation and the commentary for this work happened to have taken place.

2 Quoted in Ariel Salleh, ed. *Eco-Sufficiency & Global Justice: Women Write Political Ecology* (London: Pluto Press, 2009), 302.

3 Johan Rockström et al., "Planetary Boundaries: Exploring the Safe Operating Space for Humanity" in *Ecology and Society* 14, no. 2, at http://www.ecologyandsociety.org/vol14/iss2/art32/. For a more concise statement of the findings of the Stockholm Resilience Centre, see Johan Rockström et al., "A Safe Operating Space for Humanity" in *Nature* 461 (September 2009): 472–75.

4 Ibid.

5 Joël Cornuault, *Elisée Reclus et les Fleurs Sauvages* (Bergerac: Librairie La Brèche, 2005); Crestian Lamaison, *Elisée Reclus, l'Orthésien qui écrivait la Terre* (Orthez: Cité du Livre, 2005); Marcella Schmidt di Friedberg, ed., *Elisée Reclus. Natura e educazione* (Milan: Bruno Mondadori, 2007); and Ronald Creagh et al., eds., *Elisée Reclus, Paul Vidal de la Blache, la géographie, la cité et le monde, hier et aujourd'hui, autour de 1905* (Paris: L'Harmattan, 2009).

6 Jean-Didier Vincent, *Elisée Reclus. Géographe, anarchiste, écologiste* (Paris: Robert Laffont, 2010); Florence Deprest, *Elisée Reclus et l'Algérie colonisée* (Paris: Belin, 2012); Ronald Creagh, *Elisée Reclus et les États-Unis* (Paris: Editions Noir et Rouge, 2013); and Didier Jung, *Elisée Reclus* (Paris: Pardès, 2013).

7 Philippe Pelletier, *Géographie & anarchie. Reclus. Kropotkine. Metchnikoff.* (Paris: Editions du Monde libertaire, 2013) and *Elisée Reclus, géographie et anarchie* (Paris: Editions du Monde libertaire, 2009).

8 Federico Ferretti, *Il mondo senza la mappa: Elisée Reclus e i Geografi Anarchici* (Milan: Zero in Condotta, 2007), and *Anarchici ed editori. Reti scientifiche, editoria e lotte culturali attorno alla Nuova Geografia Universale di Elisée Reclus (1876–1894)* (Milan: Zero in Condotta, 2011).

9 Elisée Reclus, *Un nom confisqué: Elisée Reclus et sa vision des Amériques*, ed. by Ernesto Machler-Tobar (Paris: Editions INDIGO et Coté femmes, 2007); Elisée Reclus, *Projet de Globe au 100.000e*, ed. by Nikola Jankovic (Paris: Editions B2, 2011); Elisée Reclus, *Lettres de prison et d'exil*, ed. by Federico Ferretti (Lardy: A la Frontière, 2012); Elisée

Reclus, *Ecrits sociaux*, ed. by Alexandre Chollier and Federico Ferretti (Geneva: Editions Héros-limite, 2012); Elisée and Elie Reclus, *L'homme des bois, études sur les populations indiennes d'Amérique du Nord*, ed. by Alexandre Chollier and Federico Ferretti (Geneva: Editions Héros-limite, 2012).

10 For information on the Portuguese volumes and other publications, see Federico Ferretti's useful survey of recent works in "La redécouverte d'Elisée Reclus: à propos d'ouvrages récents" in *Echogéo* 21 (2012), at http://echogeo.revues.org/13173.

11 *Elisée Reclus. La passion du monde*, directed by Nicolas Eprendre and produced by Antoine Martin (Rouen, 2012).

Preface to the First Edition

1 See Elisée Reclus, *A Voyage to New Orleans: Anarchist Impressions of the Old South*, ed. and trans. John P. Clark and Camille Martin; revised and expanded edition (Enfield, N.H.: Glad Day Books, 2004).

2 Paul Reclus, "A Few Recollections on the Brothers Elie and Elisée Reclus," in *Elisée and Elie Reclus: In Memoriam*, ed. Joseph Ishill (Berkeley Heights, N.J.: Oriole Press, 1927), 5.

1 The Earth Story, the Human Story

1 Elisée Reclus, *L'Homme et la Terre* (Paris: Librairie Universelle, 1905–8), 1:i.

2 Elisée Reclus, *The Earth and Its Inhabitants: The Universal Geography*, trans. Augustus Henry Keane (London: H. Virtue, 1876–94), 1:3.

3 Elisée Reclus, *L'Homme et la Terre*, 1:114–15.

4 Ibid., 1:iv.

5 Reclus' concept of dynamic equilibrium is closer to the model of nature as a "discordant harmony" than to the simplistic idea of a "balance of nature."

6 Ibid., 1:iii–iv.

7 In this he prefigures contemporary efforts, such as Brian Swimme and Thomas Berry's account of the "Earth Story" within the larger "Universe Story." See *The Universe Story: From the Primordial Flaring Forth to the Ecozoic Era—A Celebration of the Unfolding of the Cosmos* (San Francisco: HarperCollins, 1992).

8 Reclus thus contributes to the overcoming of tendencies in Western social thought toward, on the one hand, an abstract universalism that dissolves particularity and singularity, and, on the other, a reactive and exaggerated anti-universalism that denies universality and even particularity in its cult of singularity. For an extended defense of the universal particular, see John P. Clark, *The Impossible Community: Realizing Communitarian Anarchism* (New York and London: Bloomsbury, 2013).

9 Letter to M. Roth (no specific date, 1904) in *Correspondance*, 3:285–86. Reclus paraphrases 1 John 4:20, "For anyone who does not love his brother, whom he has seen, cannot love God, whom he has not seen."

2 The Anarchist Geographer

1 *Les frères Elie et Elisée Reclus* (Paris: Les Amis d'Elisée Reclus, 1964), 17. It should be obvious that by "communism" Paul Reclus does not mean anything related to state socialism but rather a system in which the good of the community is placed above individual self-interest and obedience to conscience and principle is placed above conformity to abstract laws and regulations.

2 Ibid., 159.

3 Ibid., 162.

4 Ibid., 17.

5 Ibid., 167.

6 Ibid., 170. While the conservatism of this environment should be noted, one should not forget the strong tendency toward radicalism that was inherent in this milieu. For example, it is significant that the great libertarian theorist of the sixteenth century, Etienne de la Boétie, came out of the same Huguenot culture of southwestern France. De la Boétie, the author of the enduring anti-authoritarian classic *The Discourse of Voluntary Servitude*, was born in Sarlat in the Dordogne, the same region as Reclus. See Etienne de la Boétie, *The Politics of Obedience: The Discourse of Voluntary Servitude*, trans. Harry Kurz (New York: Free Life Editions, 1975).

7 "Développement de la liberté dans le monde," 1851 manuscript first published in *Le libertaire* (1925), quoted in *Les frères Elie et Elisée Reclus* (Paris: Les Amis d'Elisée Reclus, 1964), 50.

8 Ibid., 50.

9 Ibid., 53.

10 Ibid.

11 Ibid., 53–54.

12 For Godwin's views on "the right of private judgment," see John P. Clark, *The Philosophical Anarchism of William Godwin* (Princeton, N.J.: Princeton University Press, 1977), 136–47.

13 Elisée states that "Elie and his friends" were alone in the end (*Les frères Elie et Elisée Reclus*, 175), while Paul Reclus maintains that Elisée was in fact the only friend who remained (ibid., 23).

14 For the details of Reclus' stay in Louisiana, see Gary S. Dunbar, "Elisée Reclus in Louisiana," *Louisiana History* 23 (1982): 341–52. The article includes much fascinating material, as for example an account of Reclus' bout with yellow fever in the context of the great epidemic of 1853 (345–46).

15 See Elisée Reclus, *A Voyage to New Orleans*, ed. and trans. John P. Clark and Camille Martin (Enfield, N.H.: Glad Day Books, 2004), 51–52. While working on that book, the editors took a tour of a plantation near Félicité (which still stands but is a private home not open to visitors). We discovered that the official tour goes to great lengths to extol the grandeur of the Old South but gives no hint that a system of organized brutality ever existed there. The tour guide, a young woman dressed in antebellum garb, noted that "the slave who carried the baked pies from the kitchen outbuilding to the main house had to whistle constantly as he walked—to assure that he didn't taste the pie!" The crowd of tourists found this anecdote quite amusing, having no trouble identifying themselves with the clever planters rather than the hapless servants.

16 *Les frères Elie et Elisée Reclus*, 31.

17 Ibid. Reclus' comment is reminiscent of a famous statement by another great figure in the history of anarchist thought who was also an uncompromising opponent of slavery, Henry David Thoreau. In discussing his refusal to pay taxes to support a state that enforced slavery, Thoreau explained as follows: "How does it become a man to behave toward this American government to-day? I answer, that he cannot without disgrace be associated with it. I cannot for an instant recognize that political organization as *my* government which is the slave's government also." Henry David Thoreau, "Civil Disobedience," in *Walden and Other Writings of Henry David Thoreau*, ed. Brooks Atkinson (New York: Modern Library, 1950), 636.

18 Letter to Elie Reclus (undated) in *Correspondance* (Paris: Librairie Schleicher Frères, 1911), 1:91.

19 *Les frères Elie et Elisée Reclus*, 39.

20 W.L.G. Joerg notes that Reclus' voyage north may have taken him to Chicago and even as far northeast as Niagara Falls. He bases this view on research on Reclus' life by the great historian of anarchism Max Nettlau. See Joerg, "The Geography of North America: A History of Its Regional Exposition," *Geographical Review* 26 (1936): 648.

21 Elisée Reclus, *Voyage à la Sierra-Nevada de Saint-Marthe: Paysage de la nature tropicale* (Paris: Hachette, 1861).

22 *Les frères Elie et Elisée Reclus*, 43.

23 Peter Kropotkin, "Elisée Reclus," in *Elisée and Elie Reclus: In Memoriam*, ed. Joseph Ishill (Berkeley Heights, N.J.: Oriole Press, 1927), 63.

24 Letter to Mlle. de Gérando (January 1, 1882) in *Correspondance* (Paris: Librairie Schleicher Frères, 1911), 2:238. Marie Fleming cites Elie's observation that when Elisée was in his sixties he not only taught but also attended courses at the New University, always eager to learn from others. *The Geography of Freedom* (Montréal: Black Rose Books, 1988), 178.

25 Jean Grave, "Elisée Reclus" in *Elisée and Elie Reclus: In Memoriam*, ed. Ishill, 39. This is no small tribute to someone who was generally considered the most renowned geographer and one of the two or three most important anarchist theorists of his time.

26 Elisée Reclus, *La Terre: description des phénomènes de la vie du globe*, 2 vols. (Paris: Hachette, 1868–69); vol. 1 translated by B.B. Woodward and edited by Henry Woodward under the title *The Earth: A Descriptive History of the Phenomena of the Life of the Globe* (New York: Harper and Brothers, 1871), and vol. 2 translated by B.B. Woodward and edited by Henry Woodward under the title *The Ocean, Atmosphere, and Life: being the second series of a descriptive history of the phenomena of the life of the globe* (New York: Harper and Brothers, 1873).

27 Elisée Reclus, *Histoire d'un ruisseau* (Paris: Hachette, 1869).

28 Elisée Reclus, *Histoire d'une montagne* (Paris: Hetzel, 1880), translated by Bertha Lilly and John Lilly under the title *The History of a Mountain* (New York: Harper and Brothers, 1881).

29 Michael Bakunin, *La Polémique avec Mazzini: Ecrits et Matériaux*, part 1 of *Michel Bakounine et L'Italie 1875–82, Oeuvres Complètes de Bakounine*, ed. Arthur Lehning (Paris: Editions Champ Libre, 1973), 1:245.

30 Elisée Reclus, *The Earth and Its Inhabitants: The Universal Geography*, 19 vols., trans. Augustus Henry Keane (London: H. Virtue, 1876–94); originally published as *Nouvelle géographie universelle*, 19 vols., ed. Ernest George Ravenstein (Paris: Hachette, 1876–94). Herein, the authors refer to this work as the *New Universal Geography*, as in the French title, to preserve Reclus' conception of the work.

31 Patrick Geddes, "A Great Geographer: Elisée Reclus" in *Elisée and Elie Reclus: In Memoriam*, 155.

32 Gary S. Dunbar, *Elisée Reclus: Historian of Nature* (Hamden, Conn.: Archon Books, 1978), 95. See also Gary S. Dunbar, *The History of Geography* (Cooperstown, N.Y.: Gary S. Dunbar, 1996), especially chapters 3, 6, 11, 17, 18, and 26.

33 For extensive details on the uproar within the university after the invitation to Reclus was withdrawn, see Hem Day, ed., *Elisée Reclus en Belgique: sa vie, son activité* (Paris and Brussels: Pensée et Action, 1956). Among the documents reprinted in this work are the minutes of the students' organizations, which reported that a general assembly of the university's students voted to support Reclus, with only two dissenting votes (32).

34 Letter to Jean Grave (October 6, 1894) in *Correspondance* (Paris: Alfred Costes, 1925), 3:172.

35 Elisée Reclus, *The Earth and Its Inhabitants*, 19:vi.

36 Elisée Reclus, *L'Homme et la Terre*, 6 vols. (Paris: Librairie Universelle, 1905–8).

37 *Les frères Elie et Elisée Reclus*, 89.

38 Letter to Clara Mesnil (March 25, 1905) in *Correspondance*, 3:314.

3 The Dialectic of Nature and Culture

1 While the emphasis in the present discussion is on the relevance of Reclus' social geography to ecological thought and social theory, the considerable importance of his contribution in other areas, including physical geography and geology, should not be overlooked. Among Reclus' achievements is his early advocacy of the theory of continental drift and his defense of the view that this phenomenon is compatible with uniformitarian explanation. As early as 1868, in *The Earth*, he proposed that the planet is many times older than most contemporary theory indicated and that the continents formed a single landmass as recently as the Jurassic period. In 1979, an intriguing discussion of Reclus' geological significance appeared in the journal *Geology*. In his article "Elisée Reclus: Neglected Geologic Pioneer and First Continental Drift Advocate," James O. Berkland concludes that Reclus "was a peer of the geologic greats of the nineteenth century such as Darwin and Lyell," and that while his name "has faded to near obscurity," he "should be recognized in the history of plate tectonic theory as one of its foremost pioneers and perhaps, as its founder." See *Geology* 7 (1979): 192. In a "Comment" on this article Myrl E. Beck, Jr., suggests that Reclus' lapse into "obscurity" may have had more to do with his anarchist philosophy than with the merits of his scientific theories. See *Geology* 7 (1979): 418. In his "Reply," Berkland agrees and laments "the slow literary descent of Reclus to the status of a quasi-nonperson [*sic*]" as a case of "book-burning through neglect." In his concluding statement, Berkland surprisingly admits that "had [he] possessed full knowledge of just how 'revolutionary' Reclus really was, it is probable that [he] would not have invested the time and effort to give him well-deserved credit for his geologic accomplishments." See ibid. As geographer Gary Dunbar correctly notes, Berkland's claims are rather exaggerated. Reclus cannot be considered a "pioneer" or "founder" of continental drift theory, since others proposed the theory long before he discussed it. But while its possibility was mentioned before his time, Alfred Wegener published the first major theory of the phenomenon in 1912, and it did not receive general acceptance until the 1960s; see A. Hallam, "Alfred Wegener and the Hypothesis of Continental Drift," *Scientific American* 232, no. 2 (February 1975): 88–97. Reclus' support for continental drift thus predates its classic formulation by almost half a century and its firm establishment by almost a century. The fact that one of the foremost geographers of that era defended it at this early date was thus a significant step in the history of continental drift theory. I am grateful to Gary Dunbar and to geologist Anatol Dolgoff for their contrasting but highly enlightening views on this topic.

2 "Social ecology," in the sense used here, is an ecological philosophy based on a dialectical (but nontotalizing) view of reality. It interprets all natural and social phenomena as mutually determining parts of larger wholes and as being in a process of development and unfolding. Any whole is seen as a complex, developing unity-in-diversity that can be understood to the degree that its elements, their relationships, and the history of its development are understood. A consistent social ecology must, in view of its radically dialectical quality, reject any kind of reification of phenomena or concepts, any dualistic divisions within any sphere of reality, or any process of totalization that transforms dynamic wholes into closed systems (whether natural, social, or intellectual). These principles are applied to the evolution of humanity and the entire course of human history, in relation to natural history and the evolution of life on earth. A social ecology

is, however, more than an ecological philosophy; it is also a social practice that aims at creating a free, cooperative, ecological society in which not only the human quest to conquer nature but also all forms of domination within society are overcome.

3 Reclus' connection with contemporary social ecology has not been widely recognized. A notable exception is Peter Marshall's chapter on "Elisée Reclus: The Geographer of Liberty" in his monumental work *Demanding the Impossible: A History of Anarchism* (Oakland, CA: PM Press, 2009), 339–44. Marshall concludes that Reclus "had a profound ecological sensibility," and that he was "a forerunner of modern social ecology" (344).

4 Béatrice Giblin, "Reclus: un écologiste avant l'heure?" in *Elisée Reclus: Un géographe libertaire*, ed. Yves Lacoste, *Hérodote* 22 (1981): 110. Giblin edited and wrote the introduction for a book of selections entitled *L'Homme et la Terre—morceaux choisis* (Paris: Maspero, 1982). The entire issue of *Hérodote* containing her article is devoted to studies of Reclus' work, with a strong emphasis on the ecological implications of his social geography. Joël Cornuault, in his excellent and highly perceptive little book *Elisée Reclus, géographe et poète* (Eglise-Neuve d'Issac, France: Fédérop, 1995) also recognizes Reclus as "one of the first ecologists" and points out that although Reclus does not use the word "ecology," he showed his ecological tendencies by "commenting favorably on the works of George P. Marsh as early as 1864" (73). Cornuault shows very convincingly that Reclus' books *History of a Brook* and *History of a Mountain* are crucial to understanding his ecological sensibility and his work as a natural historian. Contemporary ecological thought has devoted little attention to the connection between geography and ecology. An exception is Thomas Berry, one of the best-known contemporary ecological thinkers, who devotes an entire chapter in one of his works to "Ecological Geography." He states that "geography provides a comprehensive context for understanding the functioning of the Earth in its larger structure," that it is "useful in appreciating the integral functioning of the various regions into which the planet is divided," and that it "provides the context for ecological thinking." *The Great Work: Our Way to the Future* (New York: Bell Tower, 1999), 86.

5 The School of Living is a radically decentralist, back-to-the-land movement based on ideas developed by Borsodi and his associates in the 1920s and early 1930s. The first experimental community was begun in 1934, and various communities and educational centers have existed ever since as part of this important grassroots movement. According to Loomis, one of the founders, the School of Living advocates a "Green Revolution" (a term used since 1940) based on such principles as "family homesteads, organic agriculture" and "activities in small communities," in addition to "freeing the land of price and speculation, cooperative credit, a stable exchange medium" and "replacing government with voluntary action." Mildred J. Loomis, *Alternative Americas* (New York: Universe Books, 1982), 73.

6 The significance of Goodman's ideas for late twentieth-century ecological thought has not been adequately appreciated. One might think in this connection of his many decentralist social and political essays, which were widely influential in the 1960s, but the philosophical basis of his contribution is found above all in his relatively early work *Gestalt Therapy* (New York: Dell, 1951), which is one of the most sophisticated theoretical works in modern anarchist thought. One of the few commentators to grasp the importance of Goodman's psychological writings and what we might call their ecological import is Taylor Stoehr. See the introduction to his Goodman collection *Nature Heals: The Psychological Essays of Paul Goodman* (New York: E.P. Dutton, 1979), xix–xxiv. Bookchin published a series of articles between 1965 and 1970 that made an early and

important contribution to the development of contemporary ecological social thought. The essays "Post-Scarcity Anarchism" and "Ecology and Revolutionary Thought" presented a radically libertarian and communitarian interpretation of ecological thinking; "Towards a Liberatory Technology" argued for the need for the development of ecological technologies; and "The Forms of Freedom" was an outline of a submerged "history of freedom" that might help inspire an alternative to the mass society of commodity consumption. These essays and others were collected in *Post-Scarcity Anarchism* (Palo Alto, Calif.: Ramparts Press, 1971) and have had a significant influence on subsequent radical ecological thought.

7 "L'Homme est la nature prenant conscience d'elle-même." Elisée Reclus, *L'Homme et la Terre* (Paris: Librairie Universelle, 1905–8), 1:1. It is instructive to compare Reclus' more ecological and dialectical concept to Marx's more environmentalist and residually dualistic conception of nature as "man's inorganic body." While the two thinkers were contemporaries (Reclus being only twelve years younger than Marx), Reclus was much more successful in transcending the spirit of the age by applying a dialectical analysis to the relationship between humanity and nature. For a discussion of Marx's philosophy of nature and his failure to develop fully the dialectical view of nature that is in fact implicit in his own thought, see John P. Clark, "Marx's Inorganic Body," *Environmental Ethics* 11 (1989): 243–58. John Bellamy Foster and Paul Burkett criticize this interpretation of Marx in "The Dialectic of Organic/Inorganic Relations: Marx and the Hegelian Philosophy of Nature," *Organization and Environment* 13 (2000): 403–25. I revise my assessment of Marx and reply to Foster and Burkett in "Marx's Natures: A Response to Foster and Burkett" *Organization and Environment* 14 (2001): 451–62.

8 Elisée Reclus, *The Ocean, Atmosphere, and Life* (New York: Harper and Brothers, 1873), 434.

9 Reclus' comprehensive, critically holistic perspective relates him to intellectual traditions beyond those that are emphasized in the present discussion. Joël Cornuault points out that Reclus' encyclopedic approach places him in the broader humanistic tradition of scholarship that preceded the extreme scientific specialization we have known since his time. He also mentions as an antecedent the Renaissance humanism of Pico della Mirandola, a reference that is perhaps surprising, but not at all inappropriate, in view of Reclus' often-eloquent affirmation of the beauty and goodness of humanity and nature. Cornuault also relates Reclus' holistic dimension to certain scientific views that were developing in the latter part of the nineteenth century, including the idea of "terrestrial unity" of Vidal de la Bache and the "principle of connection" of Brunhes. See Joël Cornuault, "'L'imagination écologique' d'Elisée Reclus: notes sur un livre de John P. Clark," *Les cahiers Elisée Reclus* 4 (1997): 2. It must always be kept in mind that the holistic dimension in Reclus has nothing to do with the positing of closed totalities, but rather concerns developing unity-in-diversity within open and relative wholes.

10 Yves Lacoste, "Editorial," in *Elisée Reclus: Un géographe libertaire*, ed. Yves Lacoste, *Hérodote* 22 (1981): 4–5.

11 Yves Lacoste, "Géographicité et géopolitique: Elisée Reclus" in *Elisée Reclus: Un géographe libertaire*, ed. Yves Lacoste, *Hérodote* 22 (1981): 14. While Reclus never had the stature in the United States that he did in France, there has been a similar process of "forgetting" in mainstream American geography. For example, if one examines *The Geographical Review* from its beginning in 1916 to the present, one finds three references to Reclus in the 1920s, three in the 1930s, two in the 1940s, and then a long silence. A modest growth of interest in Reclus among American geographers during the 1970s also parallels the situation in France and is evidenced by articles dealing with his work

in the radical geography journal *Antipode*, as well as by the publication of geographer Gary Dunbar's biography *Elisée Reclus: Historian of Nature* (Hamden, Conn.: Archon Books, 1978).

12 Elisée Reclus, "Progress," in this volume, 210.
13 Elisée Reclus, *L'Homme et la Terre*, 1:14.
14 Ibid., 1:272.
15 Elisée Reclus, "Progress," in this volume, 210.
16 Ibid., 223.
17 Ibid.
18 In Reclus' time, just as today, some views overemphasized unity and the whole and others overemphasized diversity and individual phenomena. In the past century, the organicist tradition stemming from Hegel tended toward extreme holism and social authoritarianism, while the individualist tradition arising out of classical liberalism fostered social atomism and anomic individualism. An authentically dialectical position, which might be considered a form of radical left Hegelianism, avoids both of these dangers, and interprets the whole as a dynamic, self-transcending unity-in-diversity. This is the perspective of a truly dialectical social ecology.
19 Elisée Reclus, *The Ocean, Atmosphere, and Life*, 434.
20 Elisée Reclus, *L'Homme et la Terre*, 6:254.
21 Ibid., 6:255.
22 Ibid. Reclus' holistic dimension may be compared to a similar strain in the thought of his friend and colleague Peter Kropotkin. The latter contends, for example, that geography should present a view of society and nature that will "combine in one vivid picture all separate elements of this knowledge" and "represent it as an harmonious whole, all parts of which are consequences of a few general principles and are held together by their mutual relations." *Antipode* 10, no. 3–1 (1978): 11.
23 Thérèse Dejongh, "The Brothers Reclus at the New University," in *Elisée and Elie Reclus: In Memoriam*, ed. Joseph Ishill (Berkeley Heights, N.J.: Oriole Press, 1927), 237.
24 Aldo Leopold, *A Sand County Almanac* (New York: Ballantine Books, 1970), 262, 259.
25 Edward Rothen, "Elisée Reclus' Optimism," in *Elisée and Elie Reclus: In Memoriam*, 145.
26 Ibid.
27 Ibid.
28 Elisée Reclus, *Histoire d'un ruisseau* (Arles: Actes Sud, 1995), 137.
29 This distinction was basic to Aristotle's ethics, and was important both for Hegel and for Marx's critique of Hegel on the issue of transformative praxis and the problematic of "the end of prehistory." The terms appear frequently in the literature of political ecology, including, for example in the writings of James O'Connor and other contributors to *Capitalism Nature Socialism*, the most important English-language journal of political ecology. Bookchin has made the concepts more central to his thought; however, he presents no detailed theoretical discussion of the relationship between the two realms. In his essay "Thinking Ecologically," he states that by "second nature" he means "humanity's development of a uniquely human culture, a wide variety of institutionalized human communities, an effective human technics, a richly symbolic language, and a carefully managed source of nutriment." See *The Philosophy of Social Ecology* (Montréal: Black Rose Books, 1990), 162. He describes "first nature" as the larger natural world from which second nature is "derived" and states that "the real question . . . is how second nature is derived from first nature" (163). He also posits a third natural realm, which he calls "free nature"; however, it is not an existent sphere

but rather a possibility in a future ecological society that would constitute "a nature that could reach the level of conceptual thought" (182).

30 Elisée Reclus, *L'Homme et la Terre*, 1:42.

31 Ibid.

32 Ibid., 1:117.

33 Ibid.

34 Ibid., 1:119.

35 Elisée Reclus, *The Ocean, Atmosphere, and Life*, 435.

36 One of the many similarities between the social geography of Reclus and that of Kropotkin lies in the strongly bioregional flavor of each. Myrna Breitbart points out that the latter "believed that it was necessary to reestablish a sense of community and love of place. Rootedness in a particular environment would foster greater human interaction and a more intimate relationship with one's surroundings." Myrna Breitbart, "Peter Kropotkin, Anarchist Geographer," in *Geography, Ideology and Social Concern*, ed. David Stoddart (Oxford: Blackwell, 1981), 140.

37 Ibid.

38 See Baron de Montesquieu, *The Spirit of the Laws* (New York and London: Hafner, 1949), chapters 14–17.

39 Ellsworth Huntington argues that there is "a close adjustment between life and its inorganic environment" and that factors such as "soil, climate, relief," and "position in respect to bodies of water" all "combine to form a harmonious whole" in affecting human society; *The Human Habitat* (New York: D. Van Nostrand, 1927), 16–17. It turns out that for Huntington this "harmonious whole" dictates racial hierarchy, since "racial differences" in areas such as "inherent mental capacity" are caused by the various natural factors, especially climate. See "Climate and the Evolution of Civilization," in *The Evolution of the Earth and Its Inhabitants* (New Haven, Conn.: Yale University Press, 1918), 148. Elsewhere he seeks to defend his racialist conclusions by arguing (or more accurately, speculating) that climate has had an enormous influence on inheritance through its effects on "migration, racial mixture, and natural selection," and perhaps even "mutations." See *Civilization and Climate* (New Haven, Conn.: Yale University Press, 1915), 3.

40 Elisée Reclus, *L'Homme et la Terre*, 2:91.

41 Ibid.

42 Ibid.

43 Elisée Reclus, *The Earth and Its Inhabitants: The Universal Geography*, trans. Augustus Henry Keane (London: H. Virtue, 1876–94), 1:38.

44 Ibid.

45 Elisée Reclus, "The Feeling for Nature in Modern Society," in this volume, 110.

46 Thomas Berry, "The Viable Human," in *Environmental Philosophy: From Animal Rights to Radical Ecology*, 3rd ed., Michael Zimmerman et al., eds. (Upper Saddle River, N.J.: Prentice Hall, 2001), 178.

47 Elisée Reclus, "The Feeling for Nature in Modern Society," in this volume, 110.

48 Ibid. The issue of the central role of ecological crisis in societal decline and collapse is a crucial one in world history that has been generally neglected by historians until recently. For a general overview see Clive Ponting, *A New Green History of the World: The Environment and the Collapse of Great Civilizations* (New York: Penguin Books, 2007).

49 Elisée Reclus, *Histoire d'une montagne* (Arles: Actes Sud, 1998), 224–25.

50 Elisée Reclus, *The Ocean, Atmosphere, and Life*, 526.

51 Ibid. In his view, "when man forms some loftier ideal as regards his action on the earth, he always perfectly succeeds in improving its surface, although he allows the scenery to retain its natural beauty" (527).
52 Elisée Reclus, "On Vegetarianism," in this volume, 161.
53 Elisée Reclus, *The Ocean, Atmosphere, and Life*, 527.
54 Elisée Reclus, "The Feeling for Nature in Modern Society," in this volume, 110.
55 Ibid.
56 Elisée Reclus, *Histoire d'une montagne*, 221.
57 See Wendell Berry, *The Unsettling of America* (San Francisco: Sierra Club Books, 1996). For an overview of Charbonneau's work, see John P. Clark, "Bernard Charbonneau: Regionalism and the Politics of Experience," *Capitalism Nature Socialism* 51 (2002): 41–48. For a very useful comparison of these two bioregional thinkers, see Daniel Cérézuelle, "Wendell Berry et Bernard Charbonneau, critiques de l'industrialisation de l'agriculture," in *Encyclopédie de l'Agora*, January 29, 2003, at http://agora.qc.ca/reftext.nsf/Documents/Agriculture–Wendell_Berry_et_Bernard_Charbonneau_critiques_de_par_Daniel_Cerezuelle.
58 For example, George Sessions claims that social ecologists "have yet to demonstrate an appreciation of, and commitment to, the crucial ecological importance of wilderness and biodiversity protection." See "Wilderness: Back to Basics," interview by JoAnn McAllister, in *The Trumpeter* 11 (Spring 1994): 66. Yet a dialectical position that sees humanity as "the self-consciousness of the earth," interprets history as the movement toward the realization of a "free nature," and conceives of the earth in dialectical terms as a unity-in-diversity is uniquely capable of dealing theoretically with these important issues. Steve Chase has presented a very circumspect analysis of the neglect of wilderness issues by Bookchin and many other social ecologists, and of the need for attention to these issues from a social ecological perspective. See "Whither the Radical Ecology Movement?" in *Defending the Earth: A Dialogue between Murray Bookchin and Dave Foreman*, Steve Chase, ed. (Boston: South End Press, 1991).
59 Elisée Reclus, *The Ocean, Atmosphere, and Life*, 529.
60 Ibid.
61 This raises an important issue not only for Reclus, but also for social ecology. While humanity can and ought to make a unique contribution to the emergence of greater freedom and creativity in nature, this contribution cannot presumably be limited a priori to the attainment of its own nondominating self-realization and to creative interaction with the natural milieu in a way that respects the integrity of nature. At this point in the history of the earth, one of the central ecological questions is the way in which human beings can reorganize society so that its impact on large areas of the earth can be reduced and finally minimized. A stronger conception of "nondomination" is needed: one that recognizes the need for the earth to have a sphere of ecological freedom and evolutionary creativity guided neither by human self-interest nor by human rationality.
62 Elisée Reclus, "The Progress of Mankind" in *Contemporary Review* 70 (July–December 1896): 782.
63 Elisée Reclus, *The Ocean, Atmosphere, and Life*, 526.
64 Ibid.
65 Ibid. The same idea is stated in "Progress," in this volume, 231.
66 Elisée Reclus, *The Ocean, Atmosphere, and Life*, 517–18.
67 Ibid., 518.
68 Elisée Reclus, *Histoire d'une montagne*, 134.

69 Julius Haast, Ferdinand von Hochstetter, and Oscar Peschel, *Ausland* (February 19, 1876), quoted in Elisée Reclus, *The Ocean, Atmosphere, and Life*, 519.

70 Elisée Reclus, *L'Homme et la Terre*, 5:300.

71 Ibid.

72 A century later, there are still right-wing "cornucopian" pro-natalist ideologists, but reactionary thought has generally shifted to an anti-natalist, population-control position. For an extensive critique of Garrett Hardin, perhaps the most famous and influential exponent of such a perspective, see John P. Clark, "The Tragedy of Common Sense; Part One: The Power of Myth" in *Capitalism Nature Socialism* 21:3 (2010): 35–54; and "The Tragedy of Common Sense; Part Two: From Ideology to Historical Reality" in *Capitalism Nature Socialism* 21:4 (2010): 34–49.

73 Elisée Reclus, *L'Homme et la Terre*, 5:332.

74 Ibid.

75 Ibid., 5:415.

76 Ibid., 5:416.

77 Ibid., 5:418.

78 Elisée Reclus, "On Vegetarianism," in this volume, 157.

79 Ibid., 157–58.

80 Ibid., 158.

81 Ibid., 160. Reclus' arguments constitute an eloquent defense of the humane treatment of animals, but they are far from conclusive as a proof of the moral necessity of unconditional vegetarianism. He presents an excellent case for the immorality of systems of food production that inflict continual suffering on animals and callously ignore the moral relevance of the attainment of goods by these beings and their self-realization. His critique is therefore quite pertinent to much of today's meat industry, with its factory farming and mechanized mass production. On the other hand, he does not demonstrate that no possible form of animal husbandry or hunting can be carried out in a humane manner. It is interesting that Reclus never subjects traditional hunting societies to the scathing criticism he directs toward the modern meat industry. Unfortunately, he fails to explore the possibility of morally relevant differences between the two systems.

82 Letter to Richard Heath (no specific date, 1884) in *Correspondance* (Paris: Librairie Schleicher Frères, 1911), 2:325.

83 Elisée Reclus, "The Extended Family," in this volume, 136–37.

84 Letter to Mlle. de Gérando (December. 8, 1903) in *Correspondance* (Paris: Alfred Costes, 1925), 3:267.

85 Letter to an unknown recipient, ibid., 3:323.

86 See Carol Gilligan, "Moral Orientation and Moral Development," in *Women and Moral Theory*, eds. Kay Kittay and Diana Meyers (Totowa, N.J.: Rowman and Littlefield, 1987), 19–33. According to Gilligan, "all human relationships, public and private, can be characterized both in terms of equality and in terms of attachment, and . . . both inequality and detachment constitute grounds for moral concern. Since everyone is vulnerable to both oppression and abandonment, two moral visions—one of justice and one of care—recur in human experience" (20). This important essay develops further the ethical implications of her groundbreaking work, *In a Different Voice: Psychological Theory and Women's Development* (Cambridge, Mass.: Harvard University Press, 1982).

4 A Philosophy of Progress

1 By comparison, utopian thinkers such as Blake, Fourier, and Morris sometimes appear much more an-archic, in the sense of being capable of questioning every *arche*—every "unquestionable" principle and every "historically necessary" form of domination.

2 Elisée Reclus, "Progress," in this volume, 218.

3 Elisée Reclus, *Evolution and Revolution* (London: W. Reeves, n.d.), 16. This brief pamphlet has been reprinted numerous times in many languages over the past century, and translations of it have unfortunately been the main source of knowledge of Reclus' political ideas for non-Francophone readers. While its old-fashioned rhetorical qualities perhaps once made it compelling, it lacks serious social analysis and conveys little of Reclus' enduring significance.

4 Elisée Reclus, "Progress," in this volume, 216.

5 Elisée Reclus, *The Ocean, Atmosphere, and Life* (New York: Harper and Brothers, 1873), 440.

6 Ibid., 443–44.

7 Elisée Reclus, *L'Homme et la Terre* (Paris: Librairie Universelle, 1905–8), 5:302. This statement and some of those cited previously are reminiscent of Marx's view that "where Nature is too lavish, she 'keeps him in hand, like a child in leading-strings.' She does not impose upon him any necessity to develop himself." Karl Marx, *Capital: A Critique of Political Economy* (New York: International Publishers, 1967), 1:513.

8 Elisée Reclus, *L'Homme et la Terre*, 5:302.

9 Elisée Reclus, "The Modern State," in this volume, 188.

10 Elisée Reclus, "Progress," in this volume, 216.

11 Elisée Reclus, "The Progress of Mankind," in *The Contemporary Review* 70 (July–Dec. 1896): 766. There are still enthusiastic admirers of the Aeta culture. For example, in 1998 the organization fPcN (friends of Peoples close to Nature) produced the short film *Save the Savages*, which it describes as "a story about the last free Savages in the Philippines." The film extols the virtues of the Aeta's traditional cooperative way of life and depicts its near destruction by logging and mining activities.

12 Elisée Reclus, "Progress," in this volume, 214.

13 Ibid., 216.

14 Letter to the editors of *Huelga General*, an anarchist journal in Spain (December 4, 1901) in *Correspondance*, 3:238–40.

15 Elisée Reclus, "The Feeling for Nature in Modern Society," in this volume, 111.

16 Elisée Reclus, "Progress," in this volume, 218.

17 Ibid.

18 Elisée Reclus, *L'Homme et la Terre*, 2:102.

19 Ibid.

20 Ibid., 2:103.

21 Ibid.

22 Ibid.

23 Ibid., 2:104.

24 Ibid., 3:178.

25 Ibid., 3:182–84. A very similar process, he says, was used later by Constantine to "kill" Christianity (182).

26 Ibid., 3:211–12.

27 Ibid.

28 Ibid., 3:212.

29 There are striking convergences on certain points between Reclus' thought and Buddhism in particular. For example, in his *History of a Brook*, he states in terms reminiscent of the Buddhist teaching of *anatta* that "much like a flowing river, we change at every moment; our life renews itself from minute to minute, and if we believe that we remain the same, this is nothing more than an illusion of the mind." *Histoire d'un Ruisseau* (Arles: Actes Sud, 1995), 204. The parallel with Heraclitus, the founder of Western dialectic, is also evident.

30 Letter to Paul Gsell, also published in *La Revue* (December 1, 1905), in *Correspondance*, 3:324. Reclus also at times recognizes that such an outlook is not what is popularly meant by the term "religion." See his letter to an unknown recipient (undated), ibid., 3:322–23.

31 Letter to M. Roth (no specific date, 1904), ibid., 3:286.

32 Elisée Reclus, *L'Homme et la Terre*, 3:388.

33 Ibid., 3:387.

34 Elisée Reclus, "Préface à la Conquête du pain de Pierre Kropotkine," http://kropot.free.fr/Reclus-PrefConq.htm.

35 Elisée Reclus, *Evolution and Revolution*, 14.

36 Elisée Reclus, "The Progress of Mankind," 762.

37 Elisée Reclus, "Progress," in this volume, 211.

38 Ibid., 232.

39 Ibid. Reclus' concept can be compared in some ways with the early Marx's idea of labor as the conscious self-creative activity of humanity and the expression of its "species-being."

40 Ibid., 233.

41 Ibid.

42 Indeed, an authentic holism, which must be a dialectical holism, is always a holism/anti-holism. Developing phenomena always have both a holistic and an anti-holistic moment. Radically dialectical social theory in particular stresses the fact that in a dialectical development there is always a supplement, a remainder, a surplus, that which cannot be reduced to an element of any supposed "synthesis."

43 Elisée Reclus, *L'Homme et la Terre*, 1:iii.

44 Ibid.

45 Ibid.

46 Alan Ritter, *Anarchism: A Theoretical Analysis* (Cambridge: Cambridge University Press, 1980), 3. While Ritter's introduction of this concept is very helpful, he unfortunately makes no reference to Reclus and other relevant figures but concentrates almost exclusively on the thought of Godwin, Proudhon, Bakunin, and Kropotkin. In fact, several of these thinkers had a very limited conception of the meaning of "communal individuality," and none of them contributed as much to developing such an idea as did Reclus.

47 Elisée Reclus, *The Ocean, Atmosphere, and Life*, 435.

48 Elisée Reclus, "Progress," in this volume, 225.

49 Ibid., 231.

50 Ibid.

51 Elisée Reclus, *The Earth: A Descriptive History of the Phenomena of the Life of the Globe*, trans. B.B. Woodward (New York: Harper and Brothers, 1871), 567.

52 Ibid.

53 Ibid.

54 Elisée Reclus, "Anarchy: By an Anarchist," *Contemporary Review* 45 (January–June 1884): 640.

55 Elisée Reclus, "Progress of Mankind," 75.
56 Ibid.
57 Elisée Reclus, *The Earth and Its Inhabitants: The Universal Geography*, trans. Augustus Henry Keane (London: H. Virtue, 1876–94), 19:iv.
58 Elisée Reclus, *L'Homme et la Terre*, 6:384–85.
59 Elisée Reclus, "Nouvelle proposition pour la suppression de l'ère chrétienne," in *Quelques écrits* (Paris and Brussels: Pensée et Action, 1956), 31.
60 Elisée Reclus, "Progress," in this volume, 221.
61 Ibid., 223.
62 Ibid.
63 Ibid.

5 Anarchism and Social Transformation

1 Yves Lacoste, review of *Espace et pouvoir*, by Paul Claval, and *Pour une géographie du pouvoir*, by Claude Raffestin, in *Elisée Reclus: Un géographe libertaire*, ed. Yves Lacoste, *Hérodote* 22 (1981): 157.
2 H. Roorda van Eysinga, "Avant tout anarchiste," in Hem Day, *Elisée Reclus (1830–1905): Savant et Anarchiste* (Paris and Brussels: Cahiers Pensée et Action, 1956),, quoted in Marie Fleming, *The Geography of Freedom: The Odyssey of Elisée Reclus* (Montréal: Black Rose Books, 1988), 20. It is quite appropriate that after Reclus' death his friend Jules Verne used Reclus as the model for an anarchist hero in one of his novels, *Les naufragés du "Jonathan"* (The Survivors of the *Jonathan*). Peter Costello, in his biography *Jules Verne: Inventor of Science Fiction* (New York: Charles Scribner's Sons, 1978), notes that the hero is "a philosophical anarchist and atheist called the Kaw-Djer, the Benefactor, by the natives," who after a shipwreck becomes the leader of the remote Hoste Island off the coast of South America and "finds himself having to organize a society, which he detests doing" (210). After solving numerous problems, he finally gives up power "and sets off to the even more remote island of Cape Horn, to live on the lighthouse which he has built there, to prevent further wrecks" (ibid.). Costello remarks that the character of the Benefactor seems to be based on Reclus, who had indeed tried to bring utopian anarchism to South America. Interestingly, between Hoste Island and Cape Horn Island lies L'Hermite Island, which, according to Costello, is named after an "anarchist explorer." Should this implausible story be true, the name of explorer in question would apparently mean "hermit" (*l'ermite* in French) and would be curiously reminiscent of an anarchist explorer whose name means "recluse."
3 Paul Reclus, "Biographie d'Elisée Reclus," in *Les Frères Elie et Elisée Reclus* (Paris: Les Amis d'Elisée Reclus, 1964), 51.
4 Ibid.
5 Elisée Reclus, "Préface à la Conquête du pain de Pierre Kropotkine," http://kropot.free.fr/Reclus-PrefConq.htm.
6 Ibid.
7 See John P. Clark, "What Is Anarchism?" in *The Anarchist Moment: Reflections on Culture, Nature and Power* (Montréal: Black Rose Books, 1984), 117–40.
8 Letter to M. Felix, professor at the New University of Brussels (February 1896) in *Correspondance* (Paris: Alfred Costes, 1925), 3:194.
9 Ibid.
10 The concept of "ordered anarchy" later became well known for its use in connection with various African stateless societies. See, for example, E.E. Evans-Pritchard,

"The Nuer of the Southern Sudan," in *African Political Systems*, ed. M. Fortes and E.E. Evans-Pritchard (London: Oxford University Press, 1940), 296.

11 Elisée Reclus, "Anarchy: By an Anarchist," *Contemporary Review* 45 (January–June 1884): 628. Although this phrase is often attributed to Marx, the formulation goes far back in the history of communitarian thought and was popularized by the so-called utopian socialists, especially Saint-Simon and his followers. It becomes very important for the communist anarchists, for whom an immediate movement toward distribution according to need is a central point of contention with both Marxists and "collectivist" anarchists. Reclus' own ethical communism was almost instinctual, going back to early religious influences that later conditioned his political and philosophical outlook. He expresses this outlook well in one of his letters: "Having received everything from others, I want to give everything back to them." Letter to an unknown recipient (undated) in *Correspondance* (Paris: Schleicher Frères, 1911), 3:323.

12 Ibid., 1:285.

13 Paul Reclus quotes from the stenographic record of the Congress of the League for Peace and Freedom in *Les Frères Elie et Elisée Reclus*, 56.

14 Letter to Elie Reclus (no date) in *Correspondance*, 1:285.

15 Letter to Georges Renard (June 2, 1888), ibid., 2:441–42.

16 Elisée Reclus, "Evolution, Revolution, and the Anarchist Ideal," in this volume, 138.

17 Ibid., 140.

18 Ibid., 141.

19 Ibid.

20 Ibid., 150.

21 Ibid., 152.

22 Reclus' shortcomings in this area are typical of revolutionaries of his era, and especially of those anarchists who looked to an uprising of workers and peasants as the source of coming social transformation. Bakunin is a much more extreme case of this uncritical approach, which had disastrous effects on the historical anarchist movement. See John P. Clark, "The Noble Lies of Power: Bakunin and the Critique of Ideology," in *Rights, Justice, and Community*, ed. Creighton Peden and John K. Roth (Lewiston, N.Y.: Edwin Mellen Press, 1992), 25–34, for an analysis of this strain in Bakunin's thought. I conclude there that his "unrealistic faith in revolutionary vanguards led him to overestimate the 'instinctual' revolutionary potential of the masses in the most extreme, and often dangerous manner. While there are many examples of his exaggerated hopes for various national groups, classes, and social strata, perhaps the most striking is his idealization of the bandits. His claims for the revolutionary role of brigands is based on no analysis of their actual place in society. The question of the nature of their consciousness, values, and character structures is ignored, as is the problem of how they might adapt to a cooperative order" (32–33).

23 Letter to Richard Heath (February 18, 1883) in *Correspondance*, 2:279.

24 Letter to Mlle. de Gérando (April 1, 1889), ibid., 2:447.

25 Letter to Richard Heath (June 2, 1903), ibid., 3:258.

26 Letter to Lilly Zibelin-Wilmerding (June 7, 1892), ibid., 3:118.

27 Ibid.

28 Letter to editors of *Sempre Avanti*, an Italian journal (June 28, 1892), ibid., 3:120–21.

29 A more critical anarchist view would see the anarchist terrorists as the early vanguard of the society of the spectacle who helped found a tradition of "Left Spectacularism." They were impatient with the slow evolutionary work of social and natural regeneration, which, as Reclus pointed out, is the necessary precondition for any later qualitative

revolutionary change of a liberatory nature. Instead, they adopted the tactic of the dramatic gesture, which was to catalyze in some magical way vast processes of social transformation. Left Spectacularism, though only rarely taking on a terroristic form, became the bane of the New Left of the 1960s.

30 From a letter (August 1889) found in the *Archives de la Préfecture de Police* in Paris, quoted in Marie Fleming, *The Geography of Freedom*, 151.

31 Letter to Richard Heath (no date) in *Correspondance*, 2:414–15. I am reminded of a contrasting anarchist view expressed by the libertarian writer and activist Karl Hess. In a conversation in the early 1970s, he described discussions of the problem of theft in the poor neighborhood in Washington, D.C., where he lived at the time. Some, he said, defended theft on the part of community members (even from the community's cooperative store) on the grounds that "stealing is the privilege of the poor." He argued against this view on the grounds that the consequences of theft are destructive for the community and that in reality "stealing is the privilege of the rich."

32 Ibid.

33 Letter to Jean Grave (Nov. 29, 1891), ibid., 3:97.

34 Text attached to letter to Jean Grave, ibid.

35 In one of his letters, Reclus touches on this problem in relation to individual acts of propaganda of the deed. However, he does not attempt there to offer a solution. See letter to *Sempre Avanti* (June 28, 1892), ibid., 3:121.

36 Elisée Reclus, "Anarchy: Extracts from a lecture delivered at South Place Institute, London on Monday, July 29th, 1895," in *Elisée and Elie Reclus: In Memoriam*, ed. Joseph Ishill (Berkeley Heights, N.J.: Oriole Press, 1927), 350.

37 On this subject, he agrees with Kropotkin, who saw the two greatest periods of advancement in human history to be those of the Greek polis and the medieval free cities. See Peter Kropotkin, "The State: Its Historic Role," in *Selected Writings on Anarchism and Revolution*, ed. Martin A. Miller (Cambridge, Mass.: MIT Press, 1970), 233. It should be observed that Reclus was in some ways inadequately critical in his assessment of the polis. Although the existence of such institutions as slavery and patriarchy do not negate the achievements of Greek democracy, it would be naïve to idealize that system and to neglect the ways in which the political realm was dialectically shaped by its interaction with other elements of the social whole. In a social system founded on hierarchy, domination, and exploitation, all institutions reflect the fundamental distribution of power, however mystified the effects of that power may be. After almost a century much the same uncritical approach to the polis appears in Bookchin's assessment of Greek democracy. For an analysis of this problem, see John P. Clark, "Beyond the limits of the city: A communitarian anarchist critique of libertarian municipalism," in *The Impossible Community: Realizing Communitarian Anarchism* (New York and London: Bloomsbury, 2013), especially "The Social and the Political," 261–64.

38 Elisée Reclus, *L'Homme et la Terre* (Paris: Librairie Universelle, 1905–8), 2:321.

39 Ibid., 2:335

40 Ibid., 3:519. Reclus also praises the Icelanders for maintaining "the principle of land to the peasants" in a rather equitable manner over a period of many centuries (515).

41 Ibid., 3:519. In referring to "all the inhabitants," Reclus conspicuously fails to mention the exclusion of women from these processes.

42 Ibid.

43 Peter Kropotkin makes similar points but presents a much more detailed discussion in *Mutual Aid*, in which he devotes two chapters to the medieval communes, cities, and

guilds. See chapters 5 and 6, "Mutual Aid in the Medieval City," in *Mutual Aid: A Factor of Evolution* (Boston: Porter Sargent, n.d.), 153–222.

44 Elisée Reclus, *L'Homme et la Terre*, 4:18.
45 Ibid., 4:18–19.
46 Ibid., 4:272.
47 Ibid., 4:14.
48 Ibid., 4:16.
49 Ibid.
50 Elisée Reclus, "Culture and Property," in this volume, 203.
51 Ibid.
52 Ibid., 203–4.
53 In view of his interest in municipal institutions, it is surprising that Reclus did not place more emphasis on the importance of the emergence of direct democracy in the Parisian sections during the Revolution. The classic anarchist interpretation of this chapter in the history of radical democracy appears in Kropotkin's history of the Revolution. See chapter 24, "The 'Districts' and the 'Sections' of Paris," and chapter 25, "The Sections of Paris under the New Municipal Law," in *The Great French Revolution* (Montréal: Black Rose Books, 1989), 180–94.
54 Elisée Reclus, *L'Homme et la Terre*, 5:28.
55 Ibid., 5:248. From this passage and other discussions it is clear that Reclus' municipalist ideas differ considerably from Bookchin's "libertarian municipalism" and proposals of some other social ecologists that anarchists seek offices in existing municipalities. He would find the idea of anarchists exercising political power within municipalities that are organized undemocratically and that are part of centralized nation-states to be contrary to his revolutionary position regarding social change. He would apply his criticisms of cooperatives and intentional communities to such reformist efforts (perhaps in even stronger terms), and would not be sympathetic to the argument that such electoral activity could help create a revolutionary situation in the future.
56 Ibid.
57 Elisée Reclus, "The Evolution of Cities," in this volume, 173.
58 Ibid.
59 Elisée Reclus, *L'Homme et la Terre*, 5:37–76.
60 Elisée Reclus, "The History of Cities," in this volume, 180.
61 Ibid., 182.
62 Elisée Reclus, *L'Homme et la Terre*, 1:145.
63 Elisée Reclus, "Evolution, Revolution, and the Anarchist Ideal," in this volume, 148–49.
64 Elisée Reclus, "Preface" to *La civilisation et les grands fleuves historiques*, by Léon Metchnikoff (Paris: Hachette, 1889), quoted in Marie Fleming, *The Anarchist Way to Socialism: Elisée Reclus and Nineteenth-Century European Anarchism* (London: Croom Helm, 1979), 150.
65 Elisée Reclus, "Anarchy: By an Anarchist," 637.
66 Elisée Reclus, "Evolution, Revolution, and the Anarchist Ideal," in this volume, 154.
67 Ibid.
68 Elisée Reclus, *L'Homme et la Terre*, 5:140.
69 Elisée Reclus, "Evolution, Revolution, and the Anarchist Ideal," in this volume, 153.
70 Ibid.
71 Letter to Richard Heath (November 12, 1902) in *Correspondance*, 3:250–51.
72 Ibid., 3:258–59.
73 Letter to Reclus' sister Louise (no specific date, 1859), ibid., 1:206.

74 Letter to Clara Koettlitz (April 12, 1895), ibid., 3:182.

75 Reclus and Kropotkin are similar in using communitarian and organicist terminology extensively in describing social phenomena. However, Kropotkin's depiction of the future society is sometimes more strongly organicist. For example, he states that "a new form of society is germinating" and that this society "will not be crystallized into certain unchangeable forms, but will continually modify its aspect, because it will be a living, evolving organism." See Peter Kropotkin, *Memoirs of a Revolutionist* (New York: Dover, 1971), 398–99. Furthermore, although both thinkers saw the need for the organic growth of a new society in free and loving personal relationships, Kropotkin was more optimistic than was Reclus about the organic development of cooperative institutions alongside the dominant authoritarian ones. See his two chapters on "Mutual Aid Amongst Ourselves," in *Mutual Aid: A Factor of Evolution*, 223–92. In *Memoirs of a Revolutionist*, Kropotkin criticizes socialist papers that "often have the tendency to become mere annals of complaints about existing conditions," while what is needed is "a record of those symptoms which everywhere announce the coming of a new era, the germination, the growing revolt against antiquated institutions" (418). He concludes, in a statement that might well be pondered by advocates of social change in our own day, that "it is hope, not despair, which makes successful revolutions" (ibid.).

76 Reclus is not mentioned, for example, in Joel Spring's *A Primer of Libertarian Education* (New York: Free Life Editions, 1975). In fact, one can review much of the extensive literature on libertarian, progressive, "open," and "free" education without finding any reference to his ideas.

77 Elisée Reclus, *L'Homme et la Terre*, 6:439.

78 Elisée Reclus, *The Ocean, Atmosphere, and Life* (New York: Harper and Brothers, 1873), 529.

79 Elisée Reclus, *L'Homme et la Terre*, 1:134.

80 Elisée Reclus, *The Ocean, Atmosphere, and Life*, 529.

81 Elisée Reclus, *L'Homme et la Terre*, 6:439.

82 Brian Swimme and Thomas Berry, *The Universe Story: From the Primordial Flaring Forth to the Ecozoic Era—A Celebration of the Unfolding of the Cosmos* (San Francisco: HarperCollins, 1992), 255. Elisée Reclus, *Histoire d'une montagne* (Arles: Actes Sud, 1998), 223.

83 Elisée Reclus, *L'Homme et la Terre*, 6:439.

84 Letter to Henri Roorda van Eysinga (November 4, 1897) in *Correspondance*, 3:203.

85 Elisée Reclus, *L'Homme et la Terre*, 6:440.

86 Ibid.

87 Letter to Mlle. de Gérando (October 8, 1881) in *Correspondance*, 2:235.

88 Ibid.

89 Elisée Reclus, *L'Evolution, la révolution et l'idéal anarchique* (Paris: P.V. Stock, 1898), 231.

6 The Critique of Domination

1 Elisée Reclus, "Anarchy: By an Anarchist," in *Contemporary Review* 45 (January–June 1884): 630–31.

2 Ibid.

3 Elisée Reclus, "The Modern State," in this volume, 193–94.

4 Ibid.

5 Ibid.

6 Ibid.

7 Ibid.

8 Ibid.
9 Ibid., 194.
10 Elisée Reclus, "Evolution, Revolution, and the Anarchist Ideal," in this volume, 147.
11 Ibid., 146.
12 Ibid., 146–47.
13 Elisée Reclus, "The Modern State," in this volume, 192.
14 Ibid.
15 Ibid., 196.
16 Ibid.
17 Ibid., 197.
18 Ibid., 198
19 Elisée Reclus, *L'Homme et la Terre* (Paris: Librairie Universelle, 1905–8), 5:304.
20 Elisée Reclus, "The Modern State," in this volume, 190.
21 Ibid.
22 Ibid.
23 Béatrice Giblin, "Elisée Reclus et les colonisations," in *Elisée Reclus: Un géographe libertaire*, ed. Yves Lacoste, *Hérodote* 22 (1981): 57. Reclus does not seem to consider the possibility that under certain conditions the multiplication and growth of "colonies of population" might eventually amount to imperialism and even cultural genocide.
24 Ibid., 67.
25 Elisée Reclus, *L'Homme et la Terre*, 5:118.
26 Ibid., 5:219.
27 Ibid., *L'Homme et la Terre*, 5:485.
28 The point, of course, is that Reclus anticipated Wittfogel's famous application of the concept of Oriental Despotism to the Soviet State. See Karl Wittfogel, *Oriental Despotism: A Comparative Study of Total Power* (New Haven, Conn.: Yale University Press, 1964).
29 Elisée Reclus, "The Modern State," in this volume, 199.
30 Ibid. Reclus does not use the terms, but he would certainly see such dangers in reformist calls for "social democracy" or "social welfare" policies.
31 Ibid., 200.
32 Ibid.
33 Ibid.
34 Ibid.
35 Elisée Reclus, *L'Homme et la Terre*, 6:258.
36 Ibid., 6:256.
37 Ibid.
38 Elisée Reclus, "Culture and Property," in this volume, 205.
39 Ibid.
40 Elisée Reclus, *L'Homme et la Terre*, 6:80.
41 Elisée Reclus, *A Voyage to New Orleans*, ed. and trans. John P. Clark and Camille Martin (Enfield, N.H.: Glad Day Books, 2004), 58.
42 Ibid.
43 Ibid.
44 Elisée Reclus, *L'Homme et la Terre*, 6:257.
45 Ibid.
46 Ibid.
47 Ibid., 5:287.
48 Ibid.

49 Ibid.
50 Elisée Reclus, "Progress," in this volume, 228.
51 Elisée Reclus, "The Feeling for Nature in Modern Society," in this volume, 108–9.
52 Elisée Reclus, *To My Brother the Peasant*, in this volume, 117.
53 Elisée Reclus, "Quelques mots sur la propriété," in *Almanach du Peuple pour 1873* (St Imier: Le Locle, 1873), 133.
54 Elisée Reclus, *To My Brother the Peasant*, in this volume, 117.
55 Elisée Reclus, *L'Homme et la Terre*, 6:326.
56 Elisée Reclus, *To My Brother the Peasant*, in this volume, 118.
57 *Michel Bakounine et les Conflits dans l'Internationale 1872, Bakounine: Oeuvres Complète*, ed. Arthur Lehning (Paris: Editions Champ Libre, 1975), 3:204.
58 Elisée Reclus, *L'Homme et la Terre*, 6:429.
59 Ibid., 6:430.
60 Ibid.
61 Ibid., 6:324.
62 Ibid.
63 Ibid.
64 Ibid.
65 Elisée Reclus, *A Voyage to New Orleans*, 50.
66 Elisée Reclus, *L'Homme et la Terre*, 6:106.
67 Ibid., 6:107.
68 Ibid.
69 Ibid., 6:108.
70 Peter Kropotkin, "Elisée Reclus," in *Elisée and Elie Reclus: In Memoriam*, ed. Joseph Ishill (Berkeley Heights, N.J.: Oriole Press, 1927), 60.
71 Letter to Mlle. Clara Koettlitz (April 12, 1895) in *Correspondance* (Paris: Librairie Schleicher Frères, 1911), 3:183.
72 Letter to Lilly Zibelin-Wilmerding (September, 1896?), ibid., 3:196.
73 Elisée Reclus, "Evolution, Revolution, and the Anarchist Ideal," in this volume, 144.
74 Elisée Reclus, "The Modern State," in this volume, 188–89.
75 Ibid., 189.
76 Ibid., 188.
77 Elisée Reclus, *L'Homme et la Terre*, 1:254. This theme has now become commonplace in social ecological thought. Bookchin discusses it in general terms, following Mumford, who analyzed it at much greater length and with considerably more richness of detail in his account of domestication. See Murray Bookchin, *The Ecology of Freedom: The Emergence and Dissolution of Hierarchy* (Palo Alto, Calif.: Cheshire Books, 1982), 52–54 and 57–61, and Lewis Mumford, *Technics and Human Development* (New York: Harcourt Brace Jovanovich, 1967), esp. chapter 7, "Garden, Home, and Mother," 142–62.
78 Elisée Reclus, *L'Homme et la Terre*, 1:255.
79 Ibid., 1:258.
80 Ibid.
81 Ibid.
82 Ibid.
83 Ibid., 1:270.
84 Elisée Reclus, "The Modern State," in this volume, 195.
85 Ibid.
86 Ibid.
87 Ibid.

88 Ibid.
89 Elisée Reclus, *L'Homme et la Terre*, 6:439.
90 Ibid.
91 Ibid., 6:492.
92 Ibid.
93 In a letter of 1897, he comments that "I have often spent the night in the woods or on beaches; often I have been satisfied with bread and water, and if official morality had not threatened me with imprisonment, I would not have been afraid to live in complete nudity." Letter to the editor of *La Vie Naturelle* (February 6, 1897) in *Correspondance*, 3:197.
94 Elisée Reclus, *L'Homme et la Terre*, 6:489.
95 Letter to Henri Roorda van Eysinga (March 16, 1891) in *Correspondance*, 3:90.
96 Ibid.
97 Elisée Reclus, *L'Homme et la Terre*, 6:486–87.
98 Letter to Henri Roorda van Eysinga (March 16, 1891) in *Correspondance*, 3:90.
99 See Leigh Summers, *Bound to Please: A History of the Victorian Corset* (Oxford: Berg, 2001); and Valerie Steele, *The Corset: A Cultural History* (New Haven, Conn.: Yale University Press, 2001).
100 Elisée Reclus, *L'Homme et la Terre*, 6:488.
101 Letter to Henri Roorda van Eysinga (March 16, 1891) in *Correspondance*, 3:91.
102 Elisée Reclus, *L'Homme et la Terre*, 6:489.
103 Letter to an unknown recipient (July 18, 1892) in *Correspondance*, 3:122.
104 Elisée Reclus, *L'Homme et la Terre*, 6:418.

7 The Legacy of Reclus: Liberty, Equality, Geography

1 Is it coincidental that the area of the land mass closest to the center of the image is east-central Africa? This was, of course, the place of origin of the human race.
2 Some of the most incisive analysis along these lines has been done by Joel Kovel. For his account of how racism is related to certain transhistorical aspects of human nature and to other historically situated social institutions, see *White Racism: A Psychohistory* (New York: Columbia University Press, 1984). A brilliant analysis of the interaction between capitalism, the state, and patriarchy in the formation of subjectivity in a society of domination is found in *The Age of Desire: Reflections of a Radical Psychoanalyst* (New York: Pantheon Books, 1981). He presents an analysis of the quest to transcend egoic selfhood and its relation to politics in *History and Spirit: An Inquiry into the Philosophy of Liberation* (Boston: Beacon Press, 1991), and finally, in *The Enemy of Nature* (London and New York: Zed Books, 2007) he demonstrates how the global capitalist system constitutes the ultimate obstacle to the flourishing of humanity and nature.

8 The Feeling for Nature in Modern Society

1 Jules Marcou (1824–98) was a French geologist who did extensive study of the Jura Mountains and North America. He produced geological maps of the United States, the British provinces of North America, and the world, and cofounded the Museum of Comparative Zoology.
2 Characters in *The Leatherstocking Tales*, a series of historical novels by James Fenimore Cooper.
3 The alcabala was a general sales tax established in Spain in the Middle Ages. Over the centuries, it increased from 5 percent to as much as 20 percent. It was at times perhaps

the largest single source of revenue for the crown but was notoriously unpopular and is thought to have had a detrimental effect on industry and trade.

4 Count Rumford (1753–1814), born Benjamin Thompson in Massachusetts, was a scientist, inventor, nutritionist, and social reformer who, because of British sympathies, left for Great Britain in 1776. Rumford is primarily known for his work on the nature of heat, for his improvements to fireplaces, and for playing a large role in founding the Royal Institution of Great Britain in 1800.

10 Anarchy

1 The anarchist philosopher Pierre-Joseph Proudhon (1809–65) is usually credited with popularizing the term and first associating it with a large-scale social movement.

2 In *Gargantua and Pantagruel* (Book 1, Chapter 57), Rabelais says of the inhabitants of the Abbey of Thélème that "their lives were governed not by laws, statutes, or rules, but according to their own volition and free will [*vouloir et franc arbitre*]." François Rabelais, *Gargantua et Pantagruel* (Paris: G. Jeune, 1957), 1:142.

3 One of the best known of these philosophical and literary utopias is *La Citta del Sole: Dialogo Poetico / The City of the Sun: A Poetical Dialogue*, trans. Daniel J. Donno (Berkeley: University of California Press, 1981), by Tommaso Campanella (1568–1639).

4 In July of 1830, Lafayette, recently named commandant of the National Guard, greeted the Duke of Orléans, the new king, with the words, "Here is the best of republics!" Despite these rousing words, Lafayette soon became disillusioned with the new regime and resigned his post.

5 The French working class fought for the Republic in the revolution of 1848, only to have its interests ignored by the new bourgeois regime, which nevertheless thanked the workers for their three months of misery on its behalf. The regime crushed a workers' rebellion in June of that year.

6 Hugo's words are from "A Juvénal" in *Les Châtiments, Œuvres Complètes de Victor Hugo*, vol. 4 (Paris: Hetzel-Quantin, 1882), 4:344.

7 The term "hierarchy" derives specifically from the Greek *hieratikos*, or "rule of the high priest," and ultimately from *hier*, "sacred," and *archia*, rule.

8 Timur Lenk, Tartar conqueror of southern and western Asia, who ruled Samarkand from 1369 to 1405.

9 Reclus refers to Descartes' strategy of methodological skepticism in which one begins by doubting everything, after which knowledge can be reconstituted "scientifically" and with certitude, founded on "clear and distinct" ideas.

10 Reclus says, *"le peuple choisi des Musantes."*

11 John Huss (1369?–1415) was a Bohemian religious reformer and martyr.

12 Giuseppe Ferrari (1812–76) was an Italian philosopher, historian, and political activist. He wrote *The Philosophy of Revolution* (1851) and *The History of Revolution* (1858).

13 The Icarians were followers of the French utopian socialist Etienne Cabet (1788–1856), who was influenced by the British utopian Robert Owen. Cabet's *Voyage en Icarie* (1840) inspired a large movement to create Icarian communities. A disastrous attempt was made to create a vast colony on the Red River in Texas in 1848. Later communities were established in Nauvoo, Illinois, and Corning, Iowa, with remnants surviving until 1898.

11 The Extended Family

1 Local name for several birds of the family *Cracidae*, found in Guiana. They are also called curassows.

2 The Agami Heron, *Agamia agami*, is sometimes considered the most beautiful of all New World herons. Its range is in tropical lowlands from southeast Mexico south on the Pacific slope to Ecuador and on the Atlantic slope to northern Bolivia and Amazonian Brazil. It is noted for its reclusive nature and relatively inaccessible habitat. See Emmet Reid Blake, *Manual of Neotropical Birds* (Chicago: University of Chicago Press, 1977).

3 Member of an Amerindian group of Quechua speakers, primarily in the Andean region of South America.

4 The Fazokl or Fazogli is a region in the eastern Sudan, near the border with Ethiopia. It is located in the foothills of the Abyssinian plateau and is crossed by the Blue Nile. The region was inhabited primarily by the Shangalla tribes, with later Funj and Arab immigration.

5 Reclus cites "Letters from Egypt." He is referring to *Letters from Egypt, Ethiopia, and the Peninsula of Sinai*, trans. Lenora and Joanna B. Horner (London, 1853) by Karl Richard Lepsius (1810–84), a German philologist and Egyptologist.

6 Any of several jumping rodents of the family *Dipodidae*, with long hind legs and a large tail.

7 Tribal people of the Tupian family of Central South America, including Brazilian Amazonia.

8 A pastoral people of Sudan.

9 A state of northeast India.

10 The Sindh is a region in the northwestern part of the Indian subcontinent. It is now one of the four provinces of the Islamic Republic of Pakistan.

11 2575–2150 B.C.E.

12 A wild sheep of the mountainous regions of Corsica and Sardinia.

13 Reclus says "other cetaceans"; however, the walrus is a pinnaped, not a cetacean.

14 Extinct large, long-horned wild oxen of the German forests.

15 Paul Pellisson-Fontanier (1624–93) was a French attorney and writer who was imprisoned in the Bastille.

16 Reclus says *"l'homme policé."* The French thus has a connotation of being "policed" or supervised.

12 Evolution, Revolution, and the Anarchist Ideal

1 Reclus' meaning here is not entirely clear. At the time he was writing, Poland no longer existed as a sovereign state, having been partitioned between Russia, Prussia, and Austria. The largest segment, of which Warsaw was the metropolis, consisted of the "Kingdom of Poland," which suffered under Russian domination. Presumably, "the order of Warsaw" means the autocratic imposition of order, as that through which the czarist regime suppressed seething nationalism, revolutionary movements, and student unrest in Poland.

2 Reclus is apparently referring to the illusory quality of freedom of speech and contract in a situation of vastly unequal power. In his time, the workers' alleged "free and voluntary agreement" to the conditions of labor when they accepted employment was used as a justification for strikebreaking and the destruction of labor organizations. Their "freedom" thus becomes a precondition for their misery and oppression.

3 Ferdinand Lassalle (1825–64) was a German socialist leader. He is noted for his reformist views, particularly the idea that the working class could gain control of the state by means of universal suffrage and then transform the economy into a system of workers' cooperatives. He was a major opponent of Marx in the socialist movement and was the object of extensive criticism in Marx's "Critique of the Gotha Program."

4 François Guizot (1787–1874) was a French statesman and historian.

5 Reclus was referring to a notorious event that had recently occurred in Paris. "On May 4, 1897, during peak shopping hours, a fire spread with astounding rapidity through the Bazar de la Charité, turning it into a huge inferno in which 117 people perished. In the midst of the panic that broke out at the beginning of the disaster, a number of lives were saved through acts of bravery. On the other hand, several people from high society who were there presented a sad spectacle." Roger Gonot, *Elisée Reclus: Prophète de l'Idéal Anarchique* (Pau, France: Editions Covedi, 1996), 73.

6 Juan Sebastian del Cano, the first circumnavigator of the earth. He sailed with Magellan, and after the latter's death, navigated the *Victoria* back to Spain, completing the circumnavigation in 1522.

7 The Vendôme Column was constructed to honor Napoléon I and his imperial army. The statue of the emperor atop the Column was removed during the Restoration but replaced by Louis-Philippe. Napoléon III later replaced this statue with a more imperial depiction of the emperor in a toga, which outraged republicans and radicals. After the Paris Commune was declared, it was decided to destroy the column. On April 12, 1871, Félix Pyat proposed demolition, stating that the column "was a monument of barbarism, a symbol of brute force and false glory, an affirmation of militarism, a negation of international law, a prominent insult to the conquered by their conquerors, a perpetual insult to one of the three great principles of the French Republic, fraternity." The column was destroyed on April 16. See Stewart Edwards, *The Paris Commune 1871* (New York: Quadrangle Books, 1973), 300–303.

8 Reclus refers to "les conventions scélérates," by which he means *les lois scélérates*, which consisted of repressive "emergency regulations" passed in 1894 and 1895 against the anarchists.

13 On Vegetarianism

1 Term used to refer to large meatpacking centers of the United States. It was first widely used to refer to Cincinnati and later to Chicago.

2 The La Plata Basin was a center of the *saladero* industry. *Saladeros* were slaughterhouses that bought cattle to produce jerked beef that was salted and dried in the sun.

3 See page 258, note 12.

4 Reclus refers to the retaliation by European forces against the Boxer Rebellion, an uprising against European imperialism in China in 1900. In June of that year, the conflict led to the killing of scores of Europeans in Beijing, including the German ambassador. In response, European troops went on a rampage, looting the city, slaughtering suspected Boxers and beheading prisoners in public.

5 Reclus says *"le lapin de garenne et le lapin de gouttière,"* literally, "the wild rabbit and the gutter rabbit." The latter refers ironically to the alley cat, *"le chat de gouttière,"* which was used for food by the poor.

6 Reclus' argument seems a bit confused here because he does not make it clear that he is describing processes at the phylogenetic rather than the ontogenetic level. It is certainly not always the case that the egg, fruit, or seed are produced when the organism that produces them no longer exits; however, on the level of the life of the species they represent biologically the fulfillment of the organism's reproductive role (though this generalization still ignores the function of care that is sometimes required in the case of the egg).

7 "All living things come from the egg."

14 The History of Cities

1 John Ruskin, *The Crown of Wild Olive* (New York: Thomas Y. Crowell, n.d.), 28–29.

2 Auguste Barbier (1805–82) was a satirical poet and writer, and a member of the French Academy. His poem "La cuve" is a rant against the evils and horrors of urban life. See Auguste Barbier, "La cuve," in *Iambes et poèmes* (Paris: P. Mascagna, 1840), 91–92.

3 See Victor Hugo, "A l'Arc de Triomphe" (*Les voix intérieures*), *Œuvres poétiques*, ed. Pierre Albouy (Paris: Pléiade, 1964), 1:936–48.

4 A term applied to immigrants from northern France who settled in the Dropt Valley and around Monségur after the Hundred Years' War. During the nineteenth century many *gavaches* came down from the mountains to work as "estivandiers," or seasonal workers, in wheat-producing areas.

5 Labonne, *Annuaire du Club alpin*, 1886. [Reclus' note]

6 Ingolfur Arnarsson was the first settler of Iceland. After being banished from Norway he set sail for Iceland. He brought along the posts from the high seat, or throne, of his home in Norway. On sighting land, he threw the pillars into the sea and asked the gods to wash them ashore at the appropriate spot for a settlement. He lost sight of the pillars and built a farm on the southeast coast. The posts were finally located along the coast to the west, and the settlement was moved to a spot that was given the Norse name "Reykjavik," or "Bay of Smoke," after the geothermal steam that rose there.

7 China was traditionally called "the Middle Kingdom" or "the Middle Flower" because of its supposed location at the center of the earth's surface.

8 Gobert, *le Gerotype*. [Reclus' note]

9 Later Stalingrad (1925–61), and now Volgograd.

10 J. G. Kohl, *Die geographische Lage der Hauptstädte Europas*. [Reclus' note]

11 Gomme, *Village Communities*, 48, 51; Green, *The Making of England*, 118. [Reclus' note]

12 This ancient city, now called Tell el-Farama, was one of Egypt's most important ports.

13 Cartagena de Indias is a seaport on the northern coast of Colombia. Portobello, a minor port on the eastern coast of Panama, was once a major center of the Spanish colonial empire. Reclus correctly notes that Portobello declined relative to Cartagena, but it was not because the former was directly displaced by the latter. It declined primarily because the Spanish treasure fleet system, which made it a center of exchange of silver from Peru and goods from Europe, had become obsolete by the eighteenth century. Cartagena's fortunes were affected to a much smaller degree.

14 The port of Athens.

15 Reclus overstates his point by using these particular examples. Cheyenne became a boomtown after the Union Pacific Railroad moved into Wyoming but experienced a severe decline when rail service was extended to Colorado, and Denver in particular. Carson City also experienced a boom when the Comstock Lode silver deposits were discovered but lapsed into two decades of depression when the mines were exhausted. This was followed, however, by a new period of boom with the discovery of additional gold, silver, and copper deposits in the area. Much of the history of Western boom towns is outlined in Duane A. Smith's *Rocky Mountain West: Colorado, Wyoming, and Montana, 1859–1915* (Albuquerque: University of New Mexico Press, 1992). See also Russell R. Elliot's *History of Nevada* (Lincoln: University of Nebraska Press, 1973).

16 V-shaped works, usually projecting from a fortified line.

17 Mile End Road and Whitechapel are in London's East End, noted in the nineteenth century for its poverty, crime, and industrial blight, in addition to its vibrant ethnic neighborhoods and radical politics.

18 Ch. Dufour, Bulletin de la Soc. Vaudoise des Sciences Naturelles, juin–sept. 1895, 145. [Reclus' note]

19 Emile Vandervelde, *L'Exode rural.* [Reclus' note]

20 The Garden City was an idea popularized by the town planner Sir Ebenezer Howard (1850–1928) and applied in several communities in England. The Garden City was designed to express such values as human scale, efficiency, beauty, and social cooperation. With a park and public buildings at the center, a green belt at the circumference, and extensive public space, the community was to combine the best features of urban and rural life. Howard's ideas are best known from his book *Garden Cities of To-morrow*, ed. F.J. Osborn (London: Faber and Faber, 1946). This work was first published in 1898 as *Tomorrow: A Peaceful Path to Real Reform.*

21 The name given to certain East Slavic tribes who settled in northeastern Germany during the late first millennium C.E. The name comes from the Old Slavic *po*, meaning "on the banks of" and "Laba," the Slavic name for the Elba.

22 Dr. Tetzner, *Globus*, April 7, 1900. [Reclus' note]

23 Lawrence Corthell, *Revue Scientifique*, June 27, 1896, 815. [Reclus' note]

24 Chr. Sandler, *Volks-Karten*, 1. [Reclus' note]

25 Edmond Demolins, *Les Français d'aujourd'hui*, 106, 107. [Reclus' note]

26 J. Denain-Darrays, *Questions diplomatiques et coloniales*, Feb. 1, 1903. [Reclus' note]

15 The Modern State

1 Reclus refers to Crete's civil war of 1897 between the Greeks and Muslims. Six major European powers (Germany, Austria, France, Italy, Great Britain, and Russia), in addition to Greece and Turkey, became involved in the conflict and ultimately imposed a peace agreement in conformity with their will.

2 Saint-Yves d'Alvaydra, *La mission des Juifs*, 41. [Reclus' note]

3 Reclus is punning on *taillable*, which refers both to taxing and to cutting.

4 Gustave Geoffroy, *L'Enfermé*, 51. [Reclus' note]

5 Reclus cites "Herbert Spencer, *Introduction to Social Science*, ch. V, 87." There is, however, no such title. He is apparently referring to chapter 5 of Spencer's *The Study of Sociology* (Ann Arbor: University of Michigan Press, 1961; reprint of the 1880 edition). There, Spencer comments that "agencies established to get remedies for crying evils, are liable to become agencies maintained and worked in a considerable degree, and sometimes chiefly, for the benefit of those who reap income from them" (75).

6 Louis Vignon, *La France en Algérie*. [Reclus' note]

16 Culture and Property

1 Paul Gille, *Société nouvelle*, March 1988. [Reclus' note]

2 Briot, *Etudes sur l'économie alpestre*. [Reclus' note]

3 Hiram was king of Tyre and a contemporary of David and Solomon. According to tradition, Hiram was "Grand Master of all Masons," and participated in the construction of Solomon's Temple. For this reason, he has been an important figure in the legendary history of Freemasonry.

4 Large estate.

5 Arthur Young, an English agronomist, traveled through France on the eve of the French Revolution.

17 Progress

1 Gibbon, in the original, states: "We may therefore acquiesce in the pleasing conclusion that every age of the world has increased and still increases the real wealth, the happiness, the knowledge, and perhaps the virtue of the human race." Edward Gibbon, *The*

Decline and Fall of the Roman Empire (London and New York: Everyman's Library, 1910), 3:519.

2 Havelock Ellis, *The Nineteenth Century*. [Reclus' note]

3 "Die Historie bekommt einen eigenthümlichen Reiz," *Weltgeschichte*, Neunter Theil, II, 4, 5, 6, etc. [Reclus' note]

4 Leopold von Ranke (1795–1886), perhaps the most famous German historian, is known as a founder of the modern objective school of historical study, which focused on the rigorous examination of primary sources. His social views were conservative and nationalistic.

5 M. Guyau, *Morale d'Epicure*, 153 et seq. [Reclus' note] Jean-Marie Guyau (1854–88) was French philosopher, poet, translator, and educator, known for his writings on ethics, aesthetics, religion, and various philosophical topics. He gained many admirers, including Nietzsche, before his early death.

6 Genesis I:10, 12, 18, 21, 25, 31. [Reclus' note]

7 Guyau, *Morale d'Epicure*, 157. [Reclus' note]

8 Elie Metchnikoff. *Etudes sur la nature humaine*. [Reclus' note]

9 Semper, *Die Philippinen und ihre Bewohner*; F. Blumentritt, *Versuch einer Ethnographie der Philippinen*; Ergänzungsheft zu den Pet. Mit., No. 67. [Reclus' note]

10 Reclus refers to the Philippine war for independence from the United States. The revolt began in February 1899 and lasted for almost three years. During the war, large segments of the population were slaughtered in some provinces, and entire populations of some towns were wiped out by battle and disease. This war has been systematically ignored by mainstream historians. See Howard Zinn, *A People's History of the United States* (New York: Harper and Row, 1980), 305–15.

11 Alphonse Pinard, *Bulletin de la Société de Géographie*, Dec. 1873. [Reclus' note]

12 A. Bastian, *Rechtszustände*. [Reclus' note]

13 Alfred Russel Wallace, *The Malay Archipelago: The land of the orang-utan, and the bird of paradise. A narrative of travel, with studies of man and nature* (New York: Harper and Brother, 1885), 598.

14 Islands of Asia west of New Guinea, north of Australia, south of the South China Sea; these include Indonesia, Melanesia, and often the Philippines.

15 *Unter den Kannibalen auf Borneo*. [Reclus' note]

16 Guillaume de Greef, *Sociologie générale élémentaire*, leçon XI, 39. [Reclus' note]

17 De Baer, Herbert Spencer, etc. [Reclus' note]

18 H. Drummond, *Ascent of Man*. [Reclus' note]

19 A former kingdom including Naples (with lower Italy) and Sicily; it united with the kingdom of Italy in 1861.

20 Reclus refers to several figures of his time who were associated with revolution. The first is the well-known English Romantic poet George Gordon, Lord Byron (1788–1824). In 1823, Byron sailed to Greece to devote his energies and resources to the cause of Greek independence from Turkey. Lajos Kossuth (1802–94) was the leader of the Hungarian movement for independence from Austria and the end of serfdom. He was president of the short-lived Hungarian Republic in 1859. Giuseppe Garibaldi (1807–82) was an Italian revolutionary and nationalist leader. He was major figure in Italian unification and a popular hero. Alexander Herzen (1812–70) was a Russian revolutionary, journalist, and writer. He saw the Russian peasant communes as the precursor of future socialism.

21 The "Hundred Doors" refers to the "doors" of the numerous tombs in the Theban Valley of the Kings in Egypt.

22 See Chapter VI, Book 1. [Reclus' note] Reclus refers to *L'Homme et la Terre* (Paris: Librairie Universelle, 1905–8), 1:321–54.

23 Reclus has in mind the United States and its famous doctrine of "Manifest Destiny." According to this theory, the American state was preordained by God and history to extend its dominion westward from the Atlantic to the Pacific.

24 Michel de Montaigne, *The Essays of Michel de Montaigne* (New York: Alfred A. Knopf, 1934), 1:189.

25 Ibid., 1:190.

26 Ibid., 1:189.

27 Ibid.

28 Peter Kropotkin, *The Conquest of Bread*. [Reclus' note]

29 H. Taine, *Philosophie de l'art dans les Pays-Bas*. [Reclus' note]

30 Herbert Spencer, *Social Statics*, 80. [Reclus' note]

31 Elie Metchnikoff. [Reclus' note]

32 Drummond, *Ascent of Man*, 101, 103. [Reclus' note]

33 Anna Forbes, *Insulinde: Experiences of a Naturalist's Wife in the Eastern Archipelago*. [Reclus' note]

34 "Misoneistes" are defined as "haters of innovation and change."

Bibliography

Those seeking additional primary and secondary materials on Reclus are directed to the Research on Anarchism Forum's *Elisée Reclus* collection at http://raforum.info/reclus/?lang=fr. It contains extensive materials, including an up-to-date bibliography of books and articles. Much useful material can also be found in the Elisée Reclus collection of the Anarchy Archives at http://dwardmac.pitzer.edu/Anarchist_Archives/bright/reclus/reclus.html. The French journal *Itinéraire* devoted a special issue in 1998 to Reclus that included an extensive listing of his works in French.

Bakunin, Michael. *La Polémique avec Mazzini: Ecrits et Matériaux. Part 1 of Michel Bakounine et L'Italie 1875–1882*, vol. 1 of *Oeuvres Complètes de Bakounine*. Edited by Arthur Lehning (Paris: Editions Champ Libre, 1973).

______. "Ecrit contre Marx." In *Michel Bakounine et les Conflits dans l'Internationale 1872*, vol. 3 of *Bakounine: Oeuvres Complètes*, edited by Arthur Lehning (Paris: Editions Champ Libre, 1975).

Barbier, Auguste. "La cuve." In *Iambes et poèmes* (Paris: P. Mascagna, 1840).

Beck, Myrl E., Jr. "Comment" on "Elisée Reclus: Neglected Geologic Pioneer and First Continental Drift Advocate." *Geology* 7, no. 9 (1979): 418.

Becker, Heiner, et al., eds. *Elisée Reclus, Itinéraire* 14 (1998).

Berkland, James O. "Elisée Reclus: Neglected Geologic Pioneer and First Continental Drift Advocate." *Geology* 7, no. 4 (1979): 189–92.

______. "Reply" to "Comment" by Myrl E. Beck. *Geology* 7, no. 9 (1979): 418.

Berry, Thomas. *The Great Work: Our Way to the Future* (New York: Bell Tower, 1999).

______. "The Viable Human." In *Environmental Philosophy: From Animal Rights to Radical Ecology*, edited by Michael Zimmerman et al. 3rd ed. (Upper Saddle River, N.J.: Prentice Hall, 2001).

Berry, Wendell. *The Unsettling of America* (San Francisco: Sierra Club Books, 1996).

Blake, Emmet Reid. *Manual of Neotropical Birds* (Chicago: University of Chicago Press, 1977).

Boino, Paul. "La Pensée Geographique d'Elisée Reclus." In *Elisée Reclus*, edited by Guy Henocque (St-Georges-d'Oléron, Belgium: Les Editions Libertaires and Editions Alternative Libertaire, 2002).

Bookchin, Murray. *The Ecology of Freedom: The Emergence and Dissolution of Hierarchy* (Palo Alto, Calif.: Cheshire Books, 1982).

______. *Post-Scarcity Anarchism* (Palo Alto, Calif.: Ramparts Press, 1971).

______. "Thinking Ecologically." In *The Philosophy of Social Ecology* (Montréal: Black Rose Books, 1990).

Breitbart, Myrna. "Peter Kropotkin, Anarchist Geographer." In *Geography, Ideology and Social Concern*, edited by David Stoddart (Oxford: Blackwell, 1981).

Campanella, Tommaso. *La Citta del Sole: Dialogo Poetico / The City of the Sun: A Poetical Dialogue*. Translated by Daniel J. Donno (Berkeley: University of California Press, 1981).

Cérézuelle, Daniel. "Wendell Berry et Bernard Charbonneau, critiques de l'industrialisation de l'agriculture." In *Encyclopédie de l'Agora* (January 29, 2003) at http://agora.qc.ca/documents/agriculture_biologique--wendell_berry_et_bernard_charbonneau_par_daniel_cerezuelle.

Chardak, Henriette. *Elisée Reclus, une vie: l'homme qui aimait la terre* (Paris: Stock, 1997).

Chase, Steve. "Whither the Radical Ecology Movement?" In *Defending the Earth: A Dialogue between Murray Bookchin and Dave Foreman*, edited by Steve Chase (Boston: South End Press, 1991).

Clark, John P. "Bernard Charbonneau: Regionalism and the Politics of Experience." *Capitalism Nature Socialism* 51 (2002): 41–48.

______. *The Impossible Community: Realizing Communitarian Anarchism* (New York and London: Bloomsbury, 2013).

______. "Marx's Inorganic Body." *Environmental Ethics* 11 (1989): 243–58.

______. "Marx's Natures: A Response to Foster and Burkett." *Organization and Environment* 14 (2001): 451–62.

______. "The Noble Lies of Power: Bakunin and the Critique of Ideology." In *Rights, Justice, and Community*, edited by Creighton Peden and John K. Roth (Lewiston, N.Y.: Edwin Mellen Press, 1992).

______. *The Philosophical Anarchism of William Godwin* (Princeton, N.J.: Princeton University Press, 1977).

______. "What Is Anarchism?" In *The Anarchist Moment: Reflections on Culture, Nature and Power* (Montréal: Black Rose Books, 1984).

Cornuault, Joël. *Reclus et les Fleurs Sauvages* (Bergerac: Librairie La Brèche, 2005).

______. *Elisée Reclus, géographe et poète* (Eglise-Neuve d'Issac, France: Fédérop, 1995).

______. "'L'imagination écologique' d'Elisée Reclus: notes sur un livre de John P. Clark." *Les cahiers Elisée Reclus* 4 (1997): 1–2.

Costello, Peter. *Jules Verne: Inventor of Science Fiction* (New York: Charles Scribner's Sons, 1978).

Creagh, Ronald. *Elisée Reclus et les États-Unis* (Paris: Editions Noir et Rouge, 2013).

______, et al. (eds.), *Elisée Reclus, Paul Vidal de la Blache, la géographie, la cité et le monde, hier et aujourd'hui, autour de 1905* (Paris: L'Harmattan, 2009).

Day, Hem, ed. *Elisée Reclus en Belgique: sa vie, son activité* (Paris and Bruxelles: Pensée et Action, 1956).

Dejongh, Thérèse. "The Brothers Reclus at the New University." In *Elisée and Elie Reclus: In Memoriam*, edited by Joseph Ishill (Berkeley Heights, N.J.: Oriole Press, 1927).

Deprest, Florence. *Elisée Reclus et l'Algérie colonisée* (Paris: Belin, 2012).

Dunbar, Gary S. *Elisée Reclus: Historian of Nature* (Hamden, Conn.: Archon Books, 1978).

______. "Elisée Reclus in Louisiana." *Louisiana History* 23 (1982): 341–52.
______. *The History of Geography* (Cooperstown, N.Y.: Gary S. Dunbar, 1996).
Edwards, Stewart. *The Paris Commune, 1871* (New York: Quadrangle Books, 1973).
Elliot, Russell R. *History of Nevada* (Lincoln: University of Nebraska Press, 1973).
Evans-Pritchard, E.E. "The Nuer of the Southern Sudan." In *African Political Systems*, edited by Meyer Fortes and E.E. Evans-Pritchard (London: Oxford University Press, 1940).
Ferretti, Federico. *Anarchici ed editori. Reti scientifiche, editoria e lotte culturali attorno alla Nuova Geografia Universale di Elisée Reclus (1876–1894)* (Milan: Zero in Condotta, 2011).
______. *Il mondo senza la mappa: Elisée Reclus e i Geografi Anarchici* (Milan: Zero in Condotta, 2007).
______. "La redécouverte d'Elisée Reclus: à propos d'ouvrages récents." *EchoGéo* 21 (2012) at http://echogeo.revues.org/13173.
Fleming, Marie. *The Anarchist Way to Socialism: Elisée Reclus and Nineteenth-Century European Anarchism* (London: Croom Helm, 1979).
______. *The Geography of Freedom* (Montréal: Black Rose Books, 1988).
Foster, John Bellamy, and Paul Burkett. "The Dialectic of Organic/Inorganic Relations: Marx and the Hegelian Philosophy of Nature." *Organization and Environment* 13 (2000): 403–25.
Geddes, Patrick. "A Great Geographer: Elisée Reclus." In *Elisée and Elie Reclus: In Memoriam*, edited by Joseph Ishill (Berkeley Heights, N.J.: Oriole Press, 1927).
Giblin, Béatrice. "Elisée Reclus et les colonisations." In *Elisée Reclus: Un géographe libertaire*, edited by Yves Lacoste, *Hérodote* 22 (1981): 56–79.
______. "Introduction" to *L'Homme et la Terre: morceaux choisis*, by Elisée Reclus. Paris: Maspero, 1982.
______. "Reclus: un écologiste avant l'heure?" In *Elisée Reclus: Un géographe libertaire*, edited by Yves Lacoste, *Hérodote* 22 (1981): 107–18.
Gibbon, Edward. *The Decline and Fall of the Roman Empire*, vol. 3 (London: Everyman's Library, 1910).
Gilligan, Carol. *In a Different Voice: Psychological Theory and Women's Development* (Cambridge, Mass.: Harvard University Press, 1982).
______. "Moral Orientation and Moral Development." In *Women and Moral Theory*, edited by Kay Kittay and Diana Meyers (Totowa, N.J.: Rowman & Littlefield, 1987).
Gonot, Roger. *Elisée Reclus: Prophète de l'Idéal Anarchique* (Pau, France: Editions Covedi, 1996).
Goodman, Paul. *Gestalt Therapy* (New York: Dell, 1951).
Grave, Jean. *"Elisée Reclus." In Elisée and Elie Reclus: In Memoriam*, edited by Joseph Ishill (Berkeley Heights, N.J.: Oriole Press, 1927).
Haast, Julius, Ferdinand von Hochstetter, and Oscar Peschel. *Ausland* (February 19, 1876), quoted in *Elisée Reclus, The Ocean, Atmosphere, and Life* (New York: Harper and Brothers, 1873).
Hallam, A. "Alfred Wegener and Hypothesis of Continental Drift." *Scientific American* 232, no. 2 (February 1975): 88–97.
Hénocque, Guy. "Elisée Reclus: Une Conscience Libre." In *Elisée Reclus*, edited by Guy Henocque (St-Georges-d'Oléron, Belgium: Les Editions Libertaires and Editions Alternative Libertaire, 2002).
Howard, Ebenezer. *Garden Cities of To-morrow*. Edited by F.J. Osborn (London: Faber and Faber, 1946).
Hugo, Victor. "L'Arc de Triomphe" (Les voix intérieures), in *Œuvres poétiques*, vol. 1 (Paris: Pléiade, 1964).

______. *"A Juvénal." In Les Châtiments, Œuvres Complètes de Victor Hugo*, vol. 4 (Paris: Hetzel-Quantin, 1882).

Huntington, Ellsworth. *Civilization and Climate* (New Haven, Conn.: Yale University Press, 1915).

______. "Climate and the Evolution of Civilization." In *The Evolution of the Earth and Its Inhabitants* (New Haven, Conn.: Yale University Press, 1918).

______. *The Human Habitat* (New York: D. Van Nostrand, 1927).

Ishill, Joseph, ed. *Elisée and Elie Reclus: In Memoriam* (Berkeley Heights, N.J.: Oriole Press, 1927).

Joerg, W.L.G. "The Geography of North America: A History of Its Regional Exposition." *Geographical Review* 26 (1936): 640–68.

Jones, John. "Rancid Parsley." In *New Species of Refrigerator Scum*, edited by Bob Smith, a special edition of *BioThrills Journal* 75 (1992): 45–55.

Jung, Didier. *Elisée Reclus* (Paris: Pardès, 2013).

Kovel, Joel. *The Age of Desire: Reflections of a Radical Psychoanalyst* (New York: Pantheon Books, 1981).

______. *The Enemy of Nature* (London and New York: Zed Books, 2007).

______. *History and Spirit: An Inquiry into the Philosophy of Liberation* (Boston: Beacon Press, 1991).

______. *White Racism: A Psychohistory* (New York: Columbia University Press, 1984).

Kropotkin, Peter. "Elisée Reclus." In *Elisée and Elie Reclus: In Memoriam*, edited by Joseph Ishill (Berkeley Heights, N.J.: Oriole Press, 1927).

______. *The Great French Revolution* (Montréal: Black Rose Books, 1989).

______. *Memoirs of a Revolutionist* (New York: Dover, 1971).

______. *Mutual Aid: A Factor of Evolution* (Boston: Porter Sargent, 1974).

______. *"The State: Its Historic Role." In Selected Writings on Anarchism and Revolution*, edited by Martin A. Miller (Cambridge, Mass.: MIT Press, 1970).

______. "What Geography Ought to Be." *Antipode* 10, no. 3–1 (1978): 6–15.

La Boétie, Etienne de. *The Politics of Obedience: The Discourse of Voluntary Servitude*. Translated by Harry Kurz (New York: Free Life Editions, 1975).

Lacoste, Yves. "Editorial." In *Elisée Reclus: Un géographe libertaire*, edited by Yves Lacoste, *Hérodote* 22 (1981): 4–5.

______, ed. *Elisée Reclus: Un géographe libertaire, Hérodote* 22 (1981).

______. "Géographicité et géopolitique: Elisée Reclus." In *Elisée Reclus: Un géographe libertaire*, edited by Yves Lacoste, *Hérodote* 22 (1981): 14.

______. "Review" of *Espace et pouvoir* by Paul Claval, and *Pour une géographie du pouvoir* by Claude Raffestin. *Elisée Reclus: Un géographe libertaire*, edited by Yves Lacoste, *Hérodote* 22 (1981): 154–57.

Lamaison, Crestian. *Elisée Reclus, l'Orthésien qui écrivait la Terre* (Orthez: Cité du Livre, 2005).

Leopold, Aldo. *A Sand County Almanac* (New York: Ballantine Books, 1970).

Loomis, Mildred J. *Alternative Americas* (New York: Universe Books, 1982).

Marshall, Peter. "Elisée Reclus: The Geographer of Liberty." In *Demanding the Impossible: A History of Anarchism* (Oakland, CA: PM Press, 2009).

Marx, Karl. *Capital: A Critique of Political Economy*, vol. 1 (New York: International, 1967).

Montaigne, Michel de. *The Essays of Michel de Montaigne*, vol. 1. (New York: Alfred A. Knopf, 1934).

Montesquieu, Charles de Secondat, baron de. *The Spirit of the Laws* (New York: Hafner, 1949).

Mumford, Lewis. *Technics and Human Development*, vol. 1 of *The Myth of the Machine* (New York: Harcourt Brace Jovanovich, 1967).

Pelletier, Philippe. *Géographie & anarchie. Reclus. Kropotkine. Metchnikoff.* (Paris: Editions du Monde libertaire, 2013).

______. *Elisée Reclus, géographie et anarchie* (Paris: Editions du Monde libertaire, 2009).

Ponting, Clive. *A New Green History of the World: The Environment and the Collapse of Great Civilizations* (New York: Penguin Books, 2007).

Rabelais, François. *Gargantua et Pantagruel*, vol. 1 (Paris: G. Jeune, 1957).

Reclus, Elisée. "L'Anarchie." In *Les Temps nouveaux* 18 (May 25–June 1, 1895).

______. "Anarchy: Extracts from a lecture delivered at South Place Institute, London on Monday, July 29th, 1895." In *Elisée and Elie Reclus: In Memoriam*, edited by Joseph Ishill (Berkeley Heights, N.J.: Oriole Press, 1927).

______. "Anarchy: By an Anarchist." *The Contemporary Review* 45 (January–June 1884): 627–41.

______. *Correspondance*, vols. 1 and 2 (Paris: Librairie Schleicher Frères, 1911).

______. *Correspondance*, vol. 3 (Paris: Alfred Costes, 1925).

______. "Développement de la liberté dans le monde," 1851 manuscript first published in *Le libertaire* (1925), quoted in Paul Reclus, "Biographie d'Elisée Reclus." In *Les frères Elie et Elisée Reclus* (Paris: Les Amis d'Elisée Reclus, 1964).

______. *The Earth: A Descriptive History of the Phenomena of the Life of the Globe*. Translated by B.B. Woodward and edited by Henry Woodward (New York: Harper and Brothers, 1871).

______. *The Earth and Its Inhabitants: The Universal Geography*. 19 vols. Translated by Augustus Henry Keane (London: H. Virtue, 1876–94).

______. *Ecrits sociaux*. Edited by Alexandre Chollier and Federico Ferretti (Geneva: Editions Héros-limite, 2012).

______. *Evolution and Revolution* (London: W. Reeves, n.d.).

______. *L'Evolution, la révolution et l'idéal anarchique* (Paris: Stock, 1898. Montréal: Lux Editions, 2004).

______. "The Great Kinship." In *Elisée and Elie Reclus: In Memoriam*, edited by Joseph Ishill (Berkeley Heights, N.J.: Oriole Press, 1927).

______. *Histoire d'une montagne* (Paris: Hetzel, 1880).

______. *Histoire d'une montagne* (Arles: Actes Sud, 1998).

______. *Histoire d'un ruisseau* (Paris: Hachette, 1869).

______. *Histoire d'un ruisseau* (Arles: Actes Sud, 1995).

______. *The History of a Mountain*. Translated by Bertha Lilly and John Lilly (New York: Harper and Brothers, 1881).

______ and Elie Reclus. *L'homme des bois, études sur les populations indiennes d'Amérique du Nord*. Edited by Alexandre Chollier and Federico Ferretti (Geneva: Editions Héros-limite, 2012).

______. *L'Homme et la Terre*. 6 vols. (Paris: Librairie Universelle, 1905–8).

______. *L'Homme et la Terre: morceaux choisis*. Edited by Béatrice Giblin (Paris: Maspero, 1982).

______. *Lettres de prison et d'exil*. Edited by Federico Ferretti (Lardy: A la Frontière, 2012).

______. *Nouvelle géographie universelle*. 19 vols. Edited by Ernest George Ravenstein (Paris: Hachette, 1876–94).

______. "Nouvelle proposition pour la suppression de l'ère chrétienne." In *Quelques écrits* (Paris-Bruxelles: Pensée et Action, 1956).

______. *The Ocean, Atmosphere, and Life: Being the second series of a descriptive history of the phenomena of the life of the globe*. Translated by B.B. Woodward and edited by Henry Woodward (New York: Harper and Brothers, 1873).

_______. "Préface à la Conquête du pain de Pierre Kropotkine," at http://kropot.free.fr/Reclus-PrefConq.htm.

_______. "Preface" to *La civilisation et les grands fleuves historiques by Léon Metchnikoff* (Paris: Hachette, 1889), quoted in Marie Fleming, *The Anarchist Way to Socialism: Elisée Reclus and Nineteenth-Century European Anarchism* (London: Croom Helm, 1979).

_______. "The Progress of Mankind." *The Contemporary Review* 70 (July–December 1896): 761–83.

_______. *Projet de Globe au 100.000e*. Edited by Nikola Jankovic http://raforum.info/reclus/spip.php?article401 (Paris: Editions B2, 2011).

_______."Quelques mots sur la propriété." In *Almanach du Peuple pour 1873* (St Imier: Le Locle, 1873).

_______. *La Terre: description des phénomènes de la vie du globe*. 2 vols. (Paris: Hachette, 1868–69).

_______. *Un nom confisqué: Elisée Reclus et sa vision des Amériques*. Edited by Machler-Tobar, Ernesto (Paris: Editions INDIGO et Coté femmes, 2007).

_______. "Vie d'Elie Reclus." In *Les frères Elie et Elisée Reclus* (Paris: Les Amis d'Elisée Reclus, 1964).

_______. *Voyage à la Sierra-Nevada de Saint-Marthe: Paysage de la nature tropicale*. (Paris: Hachette, 1861).

_______. *A Voyage to New Orleans: Anarchist Impressions of the Old South*, Translated and edited by John P. Clark and Camille Martin. Revised and expanded edition (Enfield, N.H.: Glad Day Books, 2004).

Reclus, Paul. "Biographie d'Elisée Reclus." In *Les frères Elie et Elisée Reclus* (Paris: Les Amis d'Elisée Reclus, 1964).

_______. "A Few Recollections on the Brothers Elie and Elisée Reclus." In *Elisée and Elie Reclus: In Memoriam*, edited by Joseph Ishill (Berkeley Heights, N.J.: Oriole Press, 1927).

Ritter, Alan. *Anarchism: A Theoretical Analysis* (Cambridge: Cambridge University Press, 1980).

Rockström. Johan, et al. "A Safe Operating Space for Humanity." *Nature* 461 (Sept. 2009): 472–75.

_______. "Planetary Boundaries: Exploring the Safe Operating Space for Humanity." *Ecology and Society* 14, no. 2 (2009): http://www.ecologyandsociety.org/vol14/iss2/art32/.

Rothen, Edward. "Elisée Reclus' Optimism." In *Elisée and Elie Reclus: In Memoriam*, edited by Joseph Ishill (Berkeley Heights, N.J.: Oriole Press, 1927).

Ruskin, John. *The Crown of Wild Olive* (New York: Thomas Y. Crowell, n.d.).

Salleh, Ariel, ed. *Eco-Sufficiency & Global Justice: Women Write Political Ecology* (London: Pluto Press, 2009).

Sarrazin, Hélène. *Elisée Reclus, ou, La passion du monde* (Paris: Découverte 1985).

Schmidt di Friedberg, Marcella, ed. *Elisée Reclus. Natura e educazione* (Milan: Bruno Mondadori, 2007).

Sessions, George. "Wilderness: Back to Basics." *The Trumpeter* 11 (Spring 1994): 65–70.

Smith, Duane A. *Rocky Mountain West: Colorado, Wyoming, and Montana, 1859–1915* (Albuquerque: University of New Mexico Press, 1992).

Spencer, Herbert. *The Study of Sociology* (Ann Arbor: University of Michigan Press, 1961; reprint of the 1880 edition).

Spring, Joel. *A Primer of Libertarian Education* (New York: Free Life Editions, 1975).

Steele, Valerie. *The Corset: A Cultural History* (New Haven, Conn.: Yale University Press, 2001).

Stoehr, Taylor. "Introduction" to *Nature Heals: The Psychological Essays of Paul Goodman*, by Paul Goodman (New York: E.P. Dutton, 1979).

Summers, Leigh. *Bound to Please: A History of the Victorian Corset* (Oxford: Berg, 2001)

Swimme, Brian, and Thomas Berry. *The Universe Story: From the Primordial Flaring Forth to the Ecozoic Era: A Celebration of the Unfolding of the Cosmos* (San Francisco: HarperCollins, 1992).

Thoreau, Henry David. "Civil Disobedience." In *Walden and Other Writings of Henry David Thoreau*, edited by Brooks Atkinson (New York: Modern Library, 1950).

Van Eysinga, H. Roorda. "Avant tout anarchiste." In Hem Day, *Elisée Reclus (1830–1905): Savant et Anarchiste* (Paris and Brussels: Cahiers Pensée et Action, 1956), quoted in Marie Fleming, *The Geography of Freedom: The Odyssey of Elisée Reclus* (Montréal: Black Rose Books, 1988).

Verne, Jules. *Les naufragés du "Jonathan"* (Paris: Union Générale d'Editions, 1978).

Vincent, Jean-Didier. *Elisée Reclus. Géographe, anarchiste, écologiste* (Paris: Robert Laffont, 2010).

Wallace, Alfred Russel. *The Malay Archipelago: The land of the orang-utan, and the bird of paradise. A narrative of travel, with studies of man and nature* (New York: Harper and Brothers, 1885).

Wittfogel, Karl. *Oriental Despotism: A Comparative Study of Total Power* (New Haven, Conn.: Yale University Press, 1964).

Woodcock, George. "Introduction" to *The Geography of Freedom: The Odyssey of Elisée Reclus* by Marie Fleming (Montréal: Black Rose Books, 1988).

Zinn, Howard. *A People's History of the United States* (New York: Harper and Row, 1980).

Index

About the Contributors

John Clark is Curtin Distinguished Professor of Humane Studies and the Professions, professor of philosophy, and a member of the environmental studies faculty at Loyola University. He has written a number of works on ecological philosophy and anarchist political theory, including, most recently, *The Impossible Community: Realizing Communitarian Anarchism* (Bloomsbury, 2013). He is completing a critical reinterpretation of social ecology entitled *Between Earth and Empire*. He writes for the journal *Capitalism Nature Socialism* and co-moderates the Research on Anarchism List. For many years he has been an activist in the anarchist, green, and bioregional movements. He is a member of the Education Workers' Union of the IWW.

Camille Martin is the author of four collections of poetry: *Looms* (2012), *Sonnets* (2010), *Codes of Public Sleep* (2007), and *Sesame Kiosk* (2001). She earned a PhD in English from Louisiana State University and an MFA in poetry from the University of New Orleans. A resident of New Orleans for many years, she now resides in Toronto. She maintains a website at http://www.camillemartin.ca and a literary blog at http://rogueembryo.com.

ABOUT PM PRESS

PM Press was founded at the end of 2007 by a small collection of folks with decades of publishing, media, and organizing experience. PM Press co-conspirators have published and distributed hundreds of books, pamphlets, CDs, and DVDs. Members of PM have founded enduring book fairs, spearheaded victorious tenant organizing campaigns, and worked closely with bookstores, academic conferences, and even rock bands to deliver political and challenging ideas to all walks of life. We're old enough to know what we're doing and young enough to know what's at stake.

We seek to create radical and stimulating fiction and non-fiction books, pamphlets, T-shirts, visual and audio materials to entertain, educate and inspire you. We aim to distribute these through every available channel with every available technology—whether that means you are seeing anarchist classics at our bookfair stalls; reading our latest vegan cookbook at the café; downloading geeky fiction e-books; or digging new music and timely videos from our website.

PM Press is always on the lookout for talented and skilled volunteers, artists, activists and writers to work with. If you have a great idea for a project or can contribute in some way, please get in touch.

PM Press
PO Box 23912
Oakland, CA 94623
www.pmpress.org

FRIENDS OF PM PRESS

These are indisputably momentous times—the financial system is melting down globally and the Empire is stumbling. Now more than ever there is a vital need for radical ideas.

In the six years since its founding—and on a mere shoestring—PM Press has risen to the formidable challenge of publishing and distributing knowledge and entertainment for the struggles ahead. With over 250 releases to date, we have published an impressive and stimulating array of literature, art, music, politics, and culture. Using every available medium, we've succeeded in connecting those hungry for ideas and information to those putting them into practice.

Friends of PM allows you to directly help impact, amplify, and revitalize the discourse and actions of radical writers, filmmakers, and artists. It provides us with a stable foundation from which we can build upon our early successes and provides a much-needed subsidy for the materials that can't necessarily pay their own way. You can help make that happen—and receive every new title automatically delivered to your door once a month—by joining as a Friend of PM Press. And, we'll throw in a free T-shirt when you sign up.

Here are your options:

- **$30 a month** Get all books and pamphlets plus 50% discount on all webstore purchases

- **$40 a month** Get all PM Press releases (including CDs and DVDs) plus 50% discount on all webstore purchases

- **$100 a month** Superstar—Everything plus PM merchandise, free downloads, and 50% discount on all webstore purchases

For those who can't afford $30 or more a month, we're introducing **Sustainer Rates** at $15, $10 and $5. Sustainers get a free PM Press T-shirt and a 50% discount on all purchases from our website.

Your Visa or Mastercard will be billed once a month, until you tell us to stop. Or until our efforts succeed in bringing the revolution around. Or the financial meltdown of Capital makes plastic redundant. Whichever comes first.

Revolution and Other Writings: A Political Reader

Gustav Landauer
edited and translated by Gabriel Kuhn

ISBN: 978-1-60486-054-2
$26.95 360 pages

"Landauer is the most important agitator of the radical and revolutionary movement in the entire country." This is how Gustav Landauer is described in a German police file from 1893. Twenty-six years later, Landauer would die at the hands of reactionary soldiers who overthrew the Bavarian Council Republic, a three-week attempt to realize libertarian socialism amidst the turmoil of post-World War I Germany. It was the last chapter in the life of an activist, writer, and mystic who Paul Avrich calls "the most influential German anarchist intellectual of the twentieth century."

This is the first comprehensive collection of Landauer writings in English. It includes one of his major works, *Revolution*, thirty additional essays and articles, and a selection of correspondence. The texts cover Landauer's entire political biography, from his early anarchism of the 1890s to his philosophical reflections at the turn of the century, the subsequent establishment of the Socialist Bund, his tireless agitation against the war, and the final days among the revolutionaries in Munich. Additional chapters collect Landauer's articles on radical politics in the US and Mexico, and illustrate the scope of his writing with texts on corporate capital, language, education, and Judaism. The book includes an extensive introduction, commentary, and bibliographical information, compiled by the editor and translator Gabriel Kuhn as well as a preface by Richard Day.

"If there were any justice in this world—at least as far as historical memory goes—then Gustav Landauer would be remembered, right along with Bakunin and Kropotkin, as one of anarchism's most brilliant and original theorists. Instead, history has abetted the crime of his murderers, burying his work in silence. With this anthology, Gabriel Kuhn has single-handedly redressed one of the cruelest gaps in Anglo-American anarchist literature: the absence of almost any English translations of Landauer."
— Jesse Cohn, author of *Anarchism and the Crisis of Representation: Hermeneutics, Aesthetics, Politics*

"Gustav Landauer was, without doubt, one of the brightest intellectual lights within the revolutionary circles of fin de siècle Europe. In this remarkable anthology, Gabriel Kuhn brings together an extensive and splendidly chosen collection of Landauer's most important writings, presenting them for the first time in English translation. With Landauer's ideas coming of age today perhaps more than ever before, Kuhn's work is a valuable and timely piece of scholarship, and one which should be required reading for anyone with an interest in radical social change."
— James Horrox, author of *A Living Revolution: Anarchism in the Kibbutz Movement*

Liberating Society from the State and Other Writings: A Political Reader

Erich Mühsam
edited by Gabriel Kuhn

ISBN: 978-1-60486-055-9
$26.95 320 pages

Erich Mühsam (1878-1934), poet, bohemian, revolutionary, is one of Germany's most renowned and influential anarchists. Born into a middle-class Jewish family, he challenged the conventions of bourgeois society at the turn of the century, engaged in heated debates on the rights of women and homosexuals, and traveled Europe in search of radical communes and artist colonies. He was a primary instigator of the ill-fated Bavarian Council Republic in 1919, and held the libertarian banner high during a Weimar Republic that came under increasing threat by right-wing forces. In 1933, four weeks after Hitler's ascension to power, Mühsam was arrested in his Berlin home. He spent the last sixteen months of his life in detention and died in the Oranienburg Concentration Camp in July 1934. Mühsam wrote poetry, plays, essays, articles, and diaries. His work unites a burning desire for individual liberation with anarcho-communist convictions, and bohemian strains with syndicalist tendencies. The body of his writings is immense, yet hardly any English translations exist. This collection presents not only *Liberating Society from the State: What is Communist Anarchism?*, Mühsam's main political pamphlet and one of the key texts in the history of German anarchism, but also some of his best-known poems, unbending defenses of political prisoners, passionate calls for solidarity with the lumpenproletariat, recollections of the utopian community of Monte Verità, debates on the rights of homosexuals and women, excerpts from his journals, and essays contemplating German politics and anarchist theory as much as Jewish identity and the role of intellectuals in the class struggle. An appendix documents the fate of Zenzl Mühsam, who, after her husband's death, escaped to the Soviet Union where she spent twenty years in Gulag camps.

"It has been remarked before how the history of the German libertarian and anarchist movement has yet to be written, and so the project to begin translation of some of the key works of Mühsam—one of the great names of German anarchism, yet virtually unknown in the English-speaking world—is most welcome. The struggles of the German working class in the early 20th century are perhaps some of the most bitter and misunderstood in European history, and it is time they were paid more attention. This book is the right place to start."
— Richard Parry, author of *The Bonnot Gang*

"We need new ideas. How about studying the ideal for which Erich Mühsam lived, worked, and died?"
— Augustin Souchy, author of *Beware Anarchist: A Life for Freedom*

Demanding the Impossible: A History of Anarchism

Peter Marshall

ISBN: 978-1-60486-064-1
$28.95 840 pages

Navigating the broad 'river of anarchy', from Taoism to Situationism, from Ranters to Punk rockers, from individualists to communists, from anarcho-syndicalists to anarcha-feminists, *Demanding the Impossible* is an authoritative and lively study of a widely misunderstood subject. It explores the key anarchist concepts of society and the state, freedom and equality, authority and power and investigates the successes and failure of the anarchist movements throughout the world. While remaining sympathetic to anarchism, it presents a balanced and critical account. It covers not only the classic anarchist thinkers, such as Godwin, Proudhon, Bakunin, Kropotkin, Reclus and Emma Goldman, but also other libertarian figures, such as Nietzsche, Camus, Gandhi, Foucault and Chomsky. No other book on anarchism covers so much so incisively.

In this updated edition, a new epilogue examines the most recent developments, including 'post-anarchism' and 'anarcho-primitivism' as well as the anarchist contribution to the peace, green and 'Global Justice' movements.

Demanding the Impossible is essential reading for anyone wishing to understand what anarchists stand for and what they have achieved. It will also appeal to those who want to discover how anarchism offers an inspiring and original body of ideas and practices which is more relevant than ever in the twenty-first century.

"Demanding the Impossible *is the book I always recommend when asked—as I often am—for something on the history and ideas of anarchism.*"
— Noam Chomsky

"Attractively written and fully referenced... bound to be the standard history."
— Colin Ward, *Times Educational Supplement*

"Large, labyrinthine, tentative: for me these are all adjectives of praise when applied to works of history, and Demanding the Impossible *meets all of them."*
— George Woodcock, *Independent*

Anarchist Seeds beneath the Snow: Left-Libertarian Thought and British Writers from William Morris to Colin Ward

David Goodway

ISBN: 978-1-60486-221-8
$24.95 420 pages

From William Morris to Oscar Wilde to George Orwell, left-libertarian thought has long been an important but neglected part of British cultural and political history. In *Anarchist Seeds beneath the Snow*, David Goodway seeks to recover and revitalize that indigenous anarchist tradition. This book succeeds as simultaneously a cultural history of left-libertarian thought in Britain and a demonstration of the applicability of that history to current politics. Goodway argues that a recovered anarchist tradition could—and should—be a touchstone for contemporary political radicals. Moving seamlessly from Aldous Huxley and Colin Ward to the war in Iraq, this challenging volume will energize leftist movements throughout the world.

"Anarchist Seeds beneath the Snow *is an impressive achievement for its rigorous scholarship across a wide range of sources, for collating this diverse material in a cogent and systematic narrative-cum-argument, and for elucidating it with clarity and flair… It is a book that needed to be written and now deserves to be read."*
—*Journal of William Morris Studies*

"Goodway outlines with admirable clarity the many variations in anarchist thought. By extending outwards to left-libertarians he takes on even greater diversity."
—Sheila Rowbotham, *Red Pepper*

"A splendid survey of 'left-libertarian thought' in this country, it has given me hours of delight and interest. Though it is very learned, it isn't dry. Goodway's friends in the awkward squad (especially William Blake) are both stimulating and comforting companions in today's political climate."
—A.N. Wilson, *Daily Telegraph*

"The history of the British anarchist movement has been little studied or appreciated outside of the movement itself. Anarchist Seeds beneath the Snow *should go a long way towards rectifying this blind spot in established labour and political history. His broad ranging erudition combined with a penetrating understanding of the subject matter has produced a fascinating, highly readable history."*
—Joey Cain, edwardcarpenterforum.org

The CNT in the Spanish Revolution Vols. 1–3

José Peirats
with an introduction by Chris Ealham

Vol. 1 **ISBN: 978-1-60486-207-2**
$28.00 432 pages

Vol. 2 **ISBN: 978-1-60486-208-9**
$22.95 312 pages

Vol. 3 **ISBN: 978-1-60486-209-6**
$22.95 296 pages

The CNT in the Spanish Revolution is the history of one of the most original and audacious, and arguably also the most far-reaching, of all the twentieth-century revolutions. It is the history of the giddy years of political change and hope in 1930s Spain, when the so-called 'Generation of '36', Peirats' own generation, rose up against the oppressive structures of Spanish society. It is also a history of a revolution that failed, crushed in the jaws of its enemies on both the reformist left and the reactionary right. José Peirats' account is effectively the official CNT history of the war, passionate, partisan but, above all, intelligent. Its huge sweeping canvas covers all areas of the anarchist experience—the spontaneous militias, the revolutionary collectives, the moral dilemmas occasioned by the clash of revolutionary ideals and the stark reality of the war effort against Franco and his German Nazi and Italian Fascist allies.

This new edition is carefully indexed in a way that converts the work into a usable tool for historians and makes it much easier for the general reader to dip in with greater purpose and pleasure.

"José Peirats' The CNT in the Spanish Revolution *is a landmark in the historiography of the Spanish Civil War . . . Originally published in Toulouse in the early 1950s, it was a rarity anxiously searched for by historians and others who gleefully pillaged its wealth of documentation. Even its republication in Paris in 1971 by the exiled Spanish publishing house, Ruedo Ibérico, though welcome, still left the book in the territory of specialists. For that reason alone, the present project to publish the entire work in English is to be applauded."*